SAFE LABORATORIES

PRINCIPLES AND PRACTICES FOR DESIGN AND REMODELING

EDITED BY
PETER C. ASHBROOK
MALCOLM M. RENFREW

CRC Press
Taylor & Francis Group
Boca Raton London New York

CRC Press is an imprint of the
Taylor & Francis Group, an **informa** business

First published 1991 by CRC Press
Taylor & Francis Group
6000 Broken Sound Parkway NW, Suite 300
Boca Raton, FL 33487-2742

Reissued 2018 by CRC Press

Library of Congress Cataloging-in-Publication Data

Ashbrook, Peter
Safe laboratories: principles and practices for design and remodeling / Peter Ashbrook, Malcolm Renfrew.
p. cm.
Includes bibliographical references and index.
ISBN 0-87371-200-5
1. Laboratories—Design and construction. 2. Laboratories—Safety measures. I. Renfrew, Malcolm. II. Title.
TH4652.A83 1990
542'.1—dc20 90-13453

A Library of Congress record exists under LC control number: 90013453

Publisher's Note
The publisher has gone to great lengths to ensure the quality of this reprint but points out that some imperfections in the original copies may be apparent.

ISBN 13: 978-1-315-89739-4 (hbk)
ISBN 13: 978-1-351-07649-4 (ebk)

Visit the Taylor & Francis Web site at http://www.taylorandfrancis.com and the
CRC Press Web site at http://www.crcpress.com

Preface

Laboratory safety has two major components: the design of safe facilities and good working practices by users. These two components are both important. This book focuses on the design component of laboratory safety. "User friendly" laboratories, by their well-thought-out design, encourage safe work practices. However, as several chapters point out, no matter how well a laboratory is designed, improper usage of laboratory facilities can always defeat the safety features designed into the laboratory.

As indicated by the title, the intent of this book is to introduce the reader to basic concepts in the design of safe laboratories. Section I presents general perspectives from three different, but each important, interested parties in the design process. General issues, such as codes, ventilation, plumbing, and chemical waste, are presented in Section II. Because ventilation is so frequently the source of problems in laboratories, the topic is explored in detail in Section III. Section IV contains chapters on both new and remodeled facilities, ranging in complexity from small business and high school laboratories to a highly complex chemical containment facility. The final section consists of a chapter that ties much of the earlier material together with the message that communication is essential throughout the project.

Problems facing architects and laboratory personnel whenever the construction or remodeling of facilities is proposed have increased in number and complexity over the years. Unfortunately, some of the involved architects will have had little experience in such projects and thus may rely too heavily on input from users for specifications. Because users tend to draw on familiarity with facilities in which they have worked previously, they often will not have included the latest concepts in efficient design with a desirable flexibility allowing for future changes in equipment and in mission. A useful approach is to arrange visits to recently constructed laboratories where friendly professionals are willing to share the results of their experience. However, time limitations may prevent in-depth discussions of what has worked and what has not. Moreover, it is not easy from the outside to identify laboratories which are models to be emulated. Hence, in addition to visitations, prudent researchers will go to the literature for expanding and updating their knowledge of new laboratory specifications.

In part, many design problems arise because there are no absolute, error-free answers to many of the questions which must be asked. Further, inadequate funding of a project frequently constrains the ability to provide for easy modification in subsequent years as user requirements change. Also, there will be different ways, each with earnest advocates, to arrive at desired goals.

In particular, problems involving safety for workers and for neighbors have loomed larger as recognition has grown that facilities and gear must provide protection not only against acute hazards such as spills, fire, and explosions, but also against exposure to low concentrations of substances with the potential for physiological damage not evident until years later. Similarly, we currently have only suspicions of difficulties ahead from exposure to the low frequency electromagnetic radiation which pervades in modern instrument rooms.

Always present is the threat of misuse of facilities by workers who may lack adequate training or may make foolish mistakes through carelessness. Good design must aim for "user friendly" facilities which will minimize any harmful results of human error.

There are different ways for attacking these safety problems, and there is room for disagreement about which way is best. For example, even in such a "simple" case as the choice of exhaust rates for fume hoods in a laboratory, we find a lack of uniformity in the recommendations of experts. There is agreement that the degree of *containment* of the materials in the process is the real test of a hood. There is growing recognition that complete containment is not feasible and that increasing the exhaust rate may not improve effectiveness. There always will be some diffusion from a hood. Experimental studies have shown that in properly used, good hood systems there is little improvement in containment over a wide range of increasing exhaust rates and that at or above some level (say about 125 fpm) turbulence will result in much poorer containment. A dozen years ago there was a common misunderstanding that OSHA was requiring 150 fpm as the exhaust rate for laboratory fume hoods which were handling regulated carcinogens. This led to a misconception that this was the proper air velocity for laboratories truly wishing to protect their workers against toxic substances and also in part to avoid liability in later suits by workers who might attribute injury to work involving the hoods.

The current OSHA laboratory standard does not carry a specific requirement for face velocities. There is a statement that exhaust rates over a range of 60 to 100 fpm will be acceptable. But there may be statutory requirements in some states. For example, California requires a minimum average face velocity of 100 fpm. There also seems to be a tacit understanding that hoods handling radioactive substances should be operated at 125 fpm (we do not know why).

Because the costs of laboratory operation increase with increases in the removal of tempered air, it is economically desirable that hood exhaust rates be set at the minimum that is judged by the users to provide acceptable levels of containment. Some differences in opinion will be noted among the authors of chapters in this book. It has been judged by the editors undesirable to insist on agreement. These differences show why persons seeking guidance in laboratory design desirably should learn the full range of existing opinions in order to make an appropriate choice for a planned building.

Many of the chapters are based on papers presented in September 1988 at a symposium in Los Angeles under the auspices of the American Chemical Society's Committee on Chemical Safety and the Division of Chemical Health and Safety.* The authors, with well-established professional reputations, do not merely include references to codes and statutes but often make personal judgments as to current best practice. The papers are regarded as contributing to education rather than prescriptions for "how to do it". References for desirable supplementary readings are included.

We are grateful to Stanley Pine, who as Chair of the American Chemical Society's Committee on Chemical Safety helped to identify this topic as one to which we could make a contribution. In addition to the obvious contributions of the authors, we wish to acknowledge the assistance of William G. Mikell, Louis J. Di Berardinis, and Paul J. Restivo in reviewing the manuscripts and providing helpful input.

* Chapters 1, 2, 3, 4, 6, 9, 10, 13, 15, 16, and 18.

The Editors

Peter C. Ashbrook received his Bachelor's Degree in Chemistry from Carleton College and his Master's Degree in Environmental Health from the University of Minnesota. He spent five years with the Minnesota State Planning Agency as a planner analyzing environmental issues of state-wide significance. For the last eight years he has headed the hazardous waste management program for the Division of Environmental Health and Safety at the University of Illinois at Urbana-Champaign. He has been active in the American Chemical Society, chairing the Facilities Subcommittee of the Committee on Chemical Safety, which organized the symposium on which this book is based. He currently serves on the ACS Task Force on RCRA. He was also a member of the hazardous waste task force for the National Association of College and University Business Officers (NACUBO).

Malcolm M. Renfrew received his B.S. and M.S. degrees from the University of Idaho and his Ph.D. in chemistry from the University of Minnesota. He was employed in research by the duPont Plastics Department, became director of chemical research and development for General Mills, and then director of research and development for Spencer Kellogg and Sons. He returned to his Alma Mater as head of the physical science division and professor of chemistry, where he continued until retirement except for one year at Stanford with the Advisory Council on College Chemistry. Throughout his career he has been active in the American Chemical Society, holding elective offices as councilor and as chairman of three divisions, including the Division of Chemical Health and Safety, for which he also served as editor of their newsletter *CHAS Notes*. He also has long served as editor of the safety column for the *Journal of Chemical Education*. Dr. Renfrew in 1985 received the award of the ACS Division of Chemical Health and Safety for outstanding contributions to science, technology, education, and communication in chemical safety. He also has been honored as an outstanding teacher by the Chemical Manufacturers Association and by the ACS Northeastern Section's Award. The Santa Clara section of ACS selected him for their Harry and Carol Mosher Award. He has been named an outstanding alumnus by the University of Idaho and by the University of Minnesota. He is a member of the American Institute of Chemical Engineers, the Society of Chemical Industry, the American Association for the Advancement of Science, the Campus Safety Association, and various other professional and academic honor societies.

Contributors

Peter C. Ashbrook
Head, Hazardous Waste Management
University of Illinois
Urbana, Illinois

Janet Baum
Payette Associates
Cambridge, Massachusetts

Leslie Bretherick
Chemical Safety Matters
Woodhayes
Dorset, England

Gerald J. Garner
Secondary Science Specialist
Los Angeles Unified School District
Van Nuys, California

E. Robinson Hoyle
(formerly) Arthur D. Little
Manchester, Massachusetts

Robert Kee
DuPont Corporation (retired)
Sagle, Idaho

Lawrence H. Keith
Radian Corporation
Austin, Texas

Michael D. Kelly
Stone, Marraccini, Patterson
San Francisco, California

Gerhard W. Knutson
Pace Laboratories
Minneapolis, Minnesota

Henry Koertge
Director, Environmental Health and Safety
University of Illinois
Urbana, Illinois

Lyle H. Phifer
Chem Service, Inc.
West Chester, Pennsylvania

Andrew T. Prokopetz
National Toxicology Program
National Institute of Environmental Health Sciences
Research Triangle Park, North Carolina

Malcolm M. Renfrew
Emeritus Professor of Chemistry
University of Idaho
Moscow, Idaho

Paul J. Restivo
Head, Environmental Health and Safety Section
Division of Environmental Health and Safety
University of Illinois
Urbana, Illinois

G. Thomas Saunders
Geneva Research
Durham, North Carolina

William Schaefer
Chemistry Division
California Institute of Technology
Pasadena, California

John Severns
Severns, Reid, and Associates
Champaign, Illinois

Richard A. Smith
Radian Corporation
Austin, Texas

Norman Steere
Minneapolis, Minnesota

R. Scott Stricoff
Arthur D. Little
Cambridge, Massachusetts

Steven Szabo
Conoco, Inc.
Ponca City, Oklahoma

Richard L. Trammell
Radian Corporation
Austin, Texas

Earl L. Walls
Earl Walls Associates
San Diego, California

Douglas B. Walters
National Toxicology Program
National Institute of Environmental Health Sciences
Research Triangle Park, North Carolina

Table of Contents

SAFE LABORATORIES

PRINCIPLES AND PRACTICES FOR DESIGN AND REMODELING

Section I: Different Perspectives on Design of Safe Laboratories

CHAPTER 1

Overview of Objectives in Design of Safe Laboratories

Leslie Bretherick

1. INTRODUCTION

The objectives in designing safety laboratories may be set out in the one sentence "to provide a workplace which is suitable for the intended purpose(s), to allow for human factors, and to include necessary intrinsic safety features, so far as is reasonably practicable."

That final phrase, "so far as is reasonably practicable", which is borrowed from the text of our 1974 Health & Safety at Work Act (U.K.), reflects the fact that at some stage in the design procedure, someone will say, "Fine, but what will all that cost?" Some degree of compromise between what is technically possible and what is economically viable is almost inevitable.

Although my definition of the overall objectives sounds simple, the means whereby these will be attained are many and complex, involving a great variety of design principles, technologies, experience, skill, and human factors, with possibly complex interactions between these. Far-sighted and skillful management of this interactive system is essential to ensure a successful design outcome.

Some of the matters which need to be considered at an early stage in design are the nature of the operations, equipment, materials, and people that will go into the laboratory, and the nature of the products and waste streams (solid, liquid, or gas) which will emerge from it. Upon these primary matters will depend the characteristics of the secondary measures, such as ventilation, services, storage, and waste disposal facilities, and emergency equipment necessary for effective operations. Points such as bench topping and floor finishes must not be forgotten.

Later chapters of this book will deal with those major features of most laboratories, ventilation and fume hoods, and also with the design features of laboratories which are intended for different purposes. I know from personal experience just how important is the full cooperation of management, architects, consulting engineers, safety officers, and laboratory users to achieve a fully integrated and effective design, and these points are also addressed in later chapters.

In this latter context, I will try to contrast two laboratory buildings in which I have worked. One of these buildings was the result of an integrated design effort and was a joy in which to work. The other was an architect's delight, designed in isolation, and a practical disaster. Although these examples are rather old now (like me), the lessons are still valid today.

2. THE FRUITS OF USER COOPERATION

The first example is an organic research building constructed in 1959. Two years before that, senior research management had insisted that laboratory users should be involved in the design procedure. At that time, I was Safety Adviser to the Research Manager, and he seconded me for a year from laboratory work to become a member of the design team and to represent interests of users.

We held meetings with laboratory staff, and a committee of users developed detailed laboratory designs with many novel safety features, some of which I will mention later. These designs were presented and discussed at a full meeting of management, architect, consulting engineers for ventilation and services, safety officer, and myself. A few modifications and compromises were necessary, but the users largely got what they wanted.

Figure 1.1 shows the general arrangement and orientation of the building, a rectangular block with five floors. The air inlet was on the prevailing wind side at second floor level, and hood exhausts discharged into a slatted "fume box" placed well above roof level and set across the prevailing wind (top arrow). The ventilation system is the first safety feature I want to mention, since the large separation of over 50 ft between fume outlets and air inlet prevented fume recirculation with very few exceptions.

The second safety feature was that each pair of laboratories had a service room between them, and one of these on each floor was specially fitted for running of unattended experiments in fume hoods overnight. Heat and smoke detectors were coupled to automatic fire extinguishers, and as in all the laboratories there was a floor gully across the doorway to prevent floods or spills from spreading. For the same reason, all pipes passing through floors went through upstanding pipe stubs sealed into the decking.

The third intrinsic safety feature was that each laboratory had a standard "safety unit" set into the wall just inside the doorway. This open front unit, about

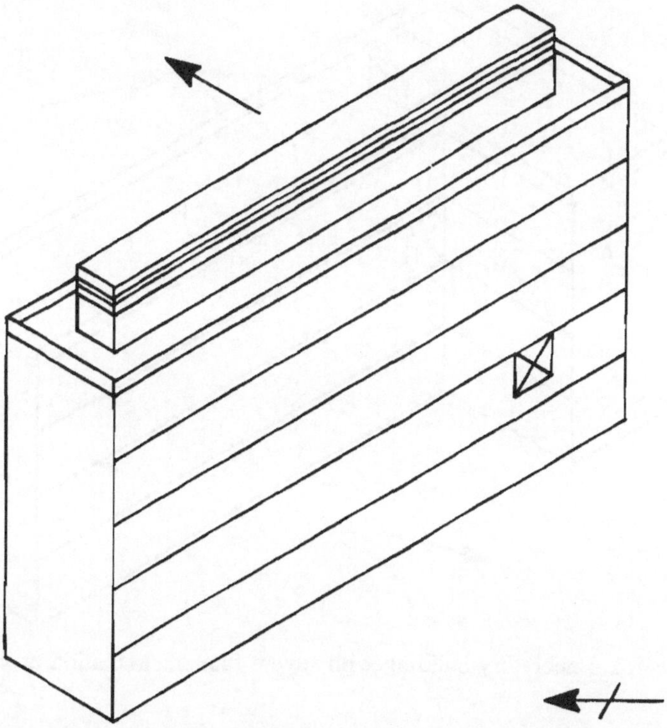

Figure 1.1. Laboratory building free of fume recirculation problems.

6′ × 2′ × 1′ deep held fire extinguisher, fire blanket, sand bucket, visor, gloves, spill kit, and other emergency items. This feature ensured that emergency equipment was immediately available in each laboratory, and that everyone knew where to find it anywhere in the building. These and many other features made it a pleasant building in which to work, and I was sorry to leave it in 1960 to relocate.

3. THE BLIGHT FROM LACK OF USER COOPERATION

The second and contrasting research building I want to discuss was completed in 1964. When I joined the company in 1962, I learned that a new building was in prospect so I set down my experiences in a diplomatic and detailed report. My manager seemed to like it, and it was passed up the line, then silence. When I later inquired about the architect's response, I was informed that he was fully experienced in all these matters and did not need advice or guidance. Anyway, it had already been decided to build a mirror-image replica of an existing building the other side of a grass area, for visual symmetry. I was already working in that other

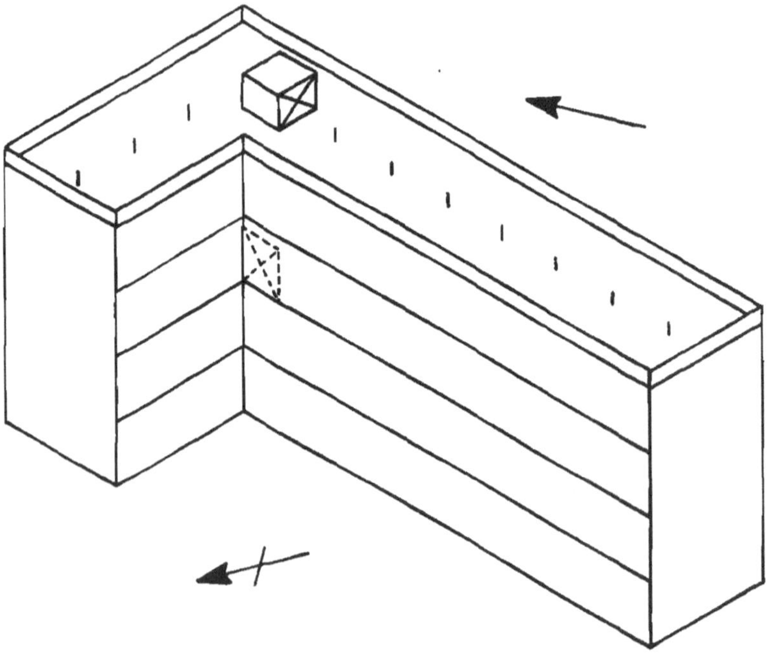

Figure 1.2. Laboratory building with severe fume recirculation problems.

building and knew of its deficiencies, but my protests were of no avail, and the mirror-image appeared in due course, complete with the known faults, and another besides.

The real problem was that external appearance counted for much more than the technical performance of the building, and this was actually confirmed on the day we moved into the building. In the corridor we met the Building Supervisor, who said, "It's a pity that you dirty chemists have to come in and foul my nice new building!"

Figure 1.2 shows the general arrangement and orientation of this L-shaped block with four floors. The air intake on the roof was cunningly located downwind of most of the fume stacks (short, so they could not be seen from the road), which absolutely guaranteed abundant fume recirculation for most of the time. Soon, discharge of noxious fumes into the hoods was prohibited, and chemical destruction or absorption in scrubbers was mandatory for a while. Later the air inlet was moved down to the third floor on the inside angle of the L, but this didn't help much, as the inlet was on the low pressure side of the building and fumes could still be pulled over the parapet and recirculated.

Then there was only one small room in the whole building which was fitted for special operations, so most experiments were run overnight in the labs, and in the absence of door gullies extensive flooding was fairly commonplace. Unsealed service pipes through floors led to multi-level floods, and on occasion a corrosive spill on the top floor got through onto a precision lathe on the ground floor.

Finally the safety equipment was not even installed until after occupation of the laboratories and emergency equipment such as fire extinguishers was fixed to any available wall space. Therefore each laboratory was completely different and nonstandard in this respect, and only the laboratory occupants would know where emergency equipment might be found. The contrast between the two buildings could not have been greater, and I was not sorry to leave the second building many years later.

The message is, therefore, **SAFETY FIRST and EARLY IN DESIGN.**

CHAPTER 2

Safe Laboratory Design: A User's Contribution

William P. Schaefer

1. INTRODUCTION

A safe laboratory can be designed and built only if people skilled in those tasks are involved. The architect, then, must know what features to include — and what pitfalls to avoid — in his layout of the building, the corridors, and the rooms. The laboratory designer must have experience in positioning equipment, laying out traffic patterns, and locating safety equipment so that these all work together to provide a safe environment for the workers. The builder must also follow the plans precisely, to produce the result intended. In reality, few architects have specific expertise in laboratory design, and most projects cannot afford a laboratory designer separate from the architect. Building contractors do not build many laboratories, so each one is a new experience for them. For all these reasons, the user of the laboratory needs to become involved from the beginning in any new laboratory construction project. The user may not have any special expertise in laboratory design and construction, but he brings an essential ingredient to the process: he knows the laboratory has to work properly. Moreover, a laboratory user can learn some of the fundamentals of laboratory design from the manuals available. The user, then, can act as an adviser to the architect, can criticize the work of a laboratory designer, and observe the progress of the building, to ensure that what was designed actually is built. Even with a skilled design and construction team involved, the user's advice is still necessary to meet the fundamental criterion of success, that the laboratory functions properly. The more closely involved the user can get with the lab designer, the architects, and the construction people, the more likely it is that the resulting building will be as good as it could be.

At the beginning someone decides that a new laboratory is needed, or an old one must be rehabilitated. Ideally that person would know how to include the necessary safety features in the design, but usually that is not the case. A lab user or lab supervisor makes the decision and can describe what must be included in the lab, and what will be done there, but the architectural features that contribute to a safe design are usually missing from the first concept. Since in a research setting laboratories are constantly changing around, being redesigned for new projects and new uses, there should be a bridge or liaison person between the final users of the laboratory, who probably are not acquainted with safe design principles, and the architects and designers who may have trouble understanding what actually will be done in a laboratory. This liaison person can help significantly in speeding laboratory design and in producing satisfied "customers". Here we will describe the stages of laboratory design and indicate how to approach it in terms of communication among all the players.

2. PREPLANNING

No one in the middle of a project wants surprises. Laboratory users do not like being told halfway through design that their hoods must be moved, or that they must give up 200 ft^2 of lab space for hazardous waste storage (they thought they would just put it out in the hall). So anyone who is even thinking about a new laboratory or lab rehabilitation needs to talk to the people who are responsible for design *right away*. These preliminary conversations often will be highly revealing, in giving the designer some ideas as to the density of construction in the lab, for example — a good clue to the first cost per square foot estimate — and in alerting the user to many facts of life in building and remodeling — the Uniform Building Code, handicap regulations, the city Fire Marshal's requirements, and so forth. By discussing all these items before any ideas have really firmed up, the surprises come when people can adjust to them more easily. They also come before any commitment of time or money has been made to the project, so that neither of these will be wasted. Of course, this is the ideal; often users will have preconceived notions and time must be spent readjusting their thinking to reality — but still, time spent early is less expensive than time spent later. What we would like to achieve from the preplanning stage is a better understanding on the part of the designer and the user as to what constraints each must operate under. At this point, people should try to avoid getting involved in detail; that is better saved for real design work. But all of the subsequent work and collaboration will be facilitated if the user and designer can get together in the preplanning stage.

3. LABORATORY DESIGN

Ideally, laboratory design would proceed in an orderly, systematic way. A user would set out the requirements for the laboratory, a designer would determine the optimum size and shape for the lab, and the chief executive would authorize the funds necessary to build it. In the normal course of events, things seldom happen this way. Certainly for rehabilitations, and often for new buildings as well, laboratories must fit into spaces that already exist, at least on paper. The user may be the most important person in the laboratory equation, but that rarely means he can dictate anything except the contents of a lab. So the orderly, systematic procedure of our fantasy becomes the iterative, political procedure of reality, and the people involved will constantly face compromises of one sort or another as a lab gets designed. This is surely the most crucial time for the laboratory users: decisions are made now that will affect them for 8, 10, or 11 hours a day over the next several (or many) years. The hope at this point is to get the actual laboratory workers — not necessarily the supervisors — talking to the designer or the architect. A good designer will have a system for interrogating the users and a checklist of items to be covered in the first interviews. Architects seldom are prepared this way and seem to rely more on the client telling them what is needed. At this point, the user or the user liaison person becomes the crucial element in the design process. He must be certain that both the utility of the laboratory and its safety are being considered, relying if need be on reading to suggest solutions to problems, or pitfalls to avoid. The designer will probably follow the rules of the applicable Building Code, but these do not give suggestions for locating hoods away from exits or aisles (necessary for safety and proper hood functioning), or provide assistance in placing the eyewash and safety showers where they are needed. So the user will have to provide this type of information to the architect, or coordinate his plans with the laboratory designer, to be sure the important ideas are being incorporated in the preliminary drawings. In any event, the first fruits of laboratory design meetings are some drawings of the floor plan, perhaps also showing utilities and elevations. An inexperienced user can be misled at this point, because the drawings always look good and their professional appearance does not invite criticism. This is just the wrong situation; here is the best opportunity for the user to really affect the final results. At this point, you should mentally "walk through" the drawings — to imagine doing routine chores in the new spaces and see if things work. Are there storage areas where they're needed? How about waste disposal? Do you need any special equipment for a procedure? Is it handy to where you use it? Are there enough sinks? water outlets? electrical outlets? Exactly what equipment will you need to plug in on this bench, in that hood? And on and on, for as many features as you can imagine. If the shape of the laboratory is still not fixed there is even more for the user to do: consider circulation patterns, aisle widths, window placement, and so forth. In a multi-room project the user needs to consider which labs need to be next to each other, which services (glass-washing, stockroom, machine shop) need to be placed where. The detail in all this can become overwhelming but it had better not be ignored. Now is the cheap time to catch mistakes, notice missing items, correct poor locations, or adjust traffic patterns. The detail in fact gets worse later, as each faucet, each drawer is specified! But at this point in design, time invested gets the greatest return. The user critiques the drawings and they are redone; the process is repeated, not *ad infinitum,* because the designer has limited time (or perhaps patience), but until there is general agreement between designer and user that things look right.

Next begins the second phase of the design work. With a conceptual design on paper, an architect (or, more likely, his draftsman) will produce the working drawings for the builder. Again the user has to look at these carefully and critically, checking the same items as previously, plus now the exact locations of all furniture, fixtures, equipment, and services. If something is missing perhaps it can still be added; real mistakes of interpretation can be fixed. In a "fast-track" project (one where construction of a building shell begins before all design work on the interior is complete), not as much can be changed at this time as is customary; too many major items have been fixed to allow the user much flexibility, except in small matters. This is the last time that fast-track or regular schedule changes can be made cheaply.

Besides checking drawings, the users are now selecting equipment — the hoods, the casework, the vacuum pumps they will use in the labs. They or the architect are specifying the lighting, the faucets, the door hardware, and other necessary but sometimes neglected items that go in the lab. These contribute to the overall safety and efficiency of the workplace, so the users must at least examine the architect's choices, to be sure that cheaper or ill-designed equipment does not compromise their laboratory's usefulness. The overriding feature of this phase is detail, detail, detail; all parties to the design are involved in endless checking and rechecking of drawings, specifications, conditions, and contracts. In fact, this process is only complete when the job is finished and the last item on the punch list is fixed; then a lucky user moves in and the designer goes on to the next job.

4. CONSTRUCTION

When the building drawings are finally approved the contractor moves in, to begin construction work. Walls and doors usually get put where you expect them, but the user has got to be alert to the possibility that *anything* on the drawing may be misinterpreted. To illustrate the problems that will arise, let us look at three examples. First, a laboratory rehabilitation project was underway and the plumber was putting in the safety showers. The drawings indicated that they go by the door but, because the doors had glass windows in them, the plumber did not think that was meant, so he moved them a foot or so off to the side, precisely over the position of a student desk that was not shown on the plumber's drawing. That mistake was caught before more than three or four showers had been installed; the user explained to the plumber that people used these showers with their clothes on.

A second mishap occurred when an electrician found he had to install a number of motor control switches for the hoods and air conditioning units. The drawing did not show where these were to go, so he consulted with the electrical engineer to do the job. Looking at their drawings, they located large open spaces on the walls in each lab and gleefully set about mounting their switch panels there. These were 3 ft wide and 5 ft high, with big gray levers and red and green lights all over them. When the user inspected the labs a day or so later it was obvious that they had taken all the space that had been reserved for centrifuges, refrigerators, and such, and made it impossible to use those walls. This was harder to fix; an area outside the labs had to be fitted up and all the switch panels moved, at some considerable cost. Both of these problems arose because individual trades do not look at drawings other than their own, and in this case the electrical and plumbing drawings did not show the moveable equipment that was shown in the architectural drawings. On top of that, the general contractor did not exercise vigilance; his superintendent should have caught both these problems, but it did not work out that way. The user generally has all the details of a job in mind and can spot problems more quickly than the contractor can.

The third problem illustrates the difficulty of reviewing drawings adequately. A plumber questioned the contractor about the installation of piping along the laboratory benches; he said it could not be done. Indeed, the draftsman had performed a real feat of legerdemain because gas and water lines, separate in two dimensions, occupied the same space in the third. It would have been impossible to put in the lines as shown and everyone had missed that on the drawings. Compounding the error in this case, the electrician had gotten in earlier and had installed his conduit and outlet boxes where the gas line was supposed to go, because the conduit fit there. The contractor should have prevented that, and if he had, would have discovered the faulty drawings early. In this case, the error was so bad no real harm was done, but in other instances misread drawings can lead to major difficulties and expense.

The lesson to be learned from these examples is to work with the general contractor, walk the job frequently, get to know the workers and talk to them about the job; find out if they see any troubles in what they're doing. In fact, that may be the most important message of all; get the people who are doing the work involved in the planning and the construction of their laboratories, and the product will be a better one.

CHAPTER 3

Designing Safe Laboratories: An Architect's Perspective*

Michael D. Kelly

1. INTRODUCTION

A discussion on the design of safe laboratories has an inherent fundamental focus on all those issues which comprise the technical details of a laboratory's function and operation. Of necessity, these details provide for the basic mechanics of how the building works, including hood design, its attendant ventilation systems, operating procedures, emergency provisions, equipment, and the like.

2. ARCHITECTURE AND "A SENSE OF COMMUNITY"

What is not readily apparent through technical analysis is the impact the environment, which houses the research functions, has upon safety and, morever, how it can enhance productivity. As architects for institutional and corporate research buildings, it is incumbent that we share our experience in an effort to describe the benefits that a commitment to a complete design processs brings to every owner and to every user of such a facility.

As architects, we are concerned with the total environment and the effect it has upon the research community. "Community" in this context is the operative function and is key to the ultimate success of a facility. The architect's primary design goal involves the resolution of the research facility as a type of institution.

*Paper presented to the American Chemical Society, Los Angeles, CA, September 28, 1988.

We are most concerned with the participants of this institution, their social order, their boundaries, their views of both the world outside and within their own institution. Such an assembly of specialists tends to be cellular and introspective. It is our responsibility, through our architecture, to enlarge and expand this perspective; the best resources genuinely available to these specialists are their fellow workers. To be successful, a research facility must foster a sense of community among its members. The structured, but spontaneous, social interface necessary for such a community is envisioned as an essential programmatic requirement, not an option nor just a desirable addition.

As an example, the first such application of this programmatic requirement begins, as you might suspect, at the front door. The imagery conveys a message to the users, reinforced daily, as to how they should think of themselves in the context of their professional activity. Properly planned, it is the initial opportunity to reinforce their own community awareness, contextually bringing together their colleagues at the start of the day's activity. The subtlety of this message is reinforced by repetition over time. Throughout a facility there must be additional areas which are perceived as opportunities to continually reinforce this social order. These are zones of confluence which must become an integral part of the facilities program and fully acknowledged as an equal partner in the planning process.

3. FLEXIBILITY'S IMPACT ON SAFETY

The architect must be dedicated to constructing the most ideal, the most flexible house for this community, expressing the workings of a research facility in its imagery. The integration of programmatic, structural, mechanical, and electrical networks comprises a literal machine, the largest component of the instruments of research within which specialists work. It is the assemblage of parts, the overlap of the various systems that drives the architecture, its function and environment. Key to long-term safety within the laboratory is the responsiveness with which the building can rationally adapt to change.

The basic planning tool for facilitating future modifications is the modularity of the initial design (Figure 3.1). We promote a generic approach to laboratory design. One, which has as its basis, the determination of an acceptable planning module(s). The repetitive and disciplined application of this module, including the planning of all required building systems, can provide for the safety mandated for the initial as well as future research programs and their users.

With rapid expansion of both technology and fields of research, R&D facilities require the flexibility to easily adapt to new programs and their associated criteria with a minimum of effort and disruption to continuing experimentation.

Architectural features, including laboratory size, partitioning, access, and functional relationships must be capable of modification. Ventilation, plumbing, and electrical systems must be made easily available in an organized manner.

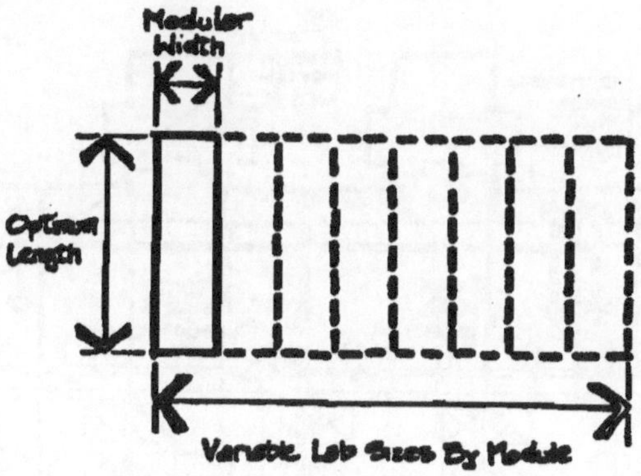

Figure 3.1.

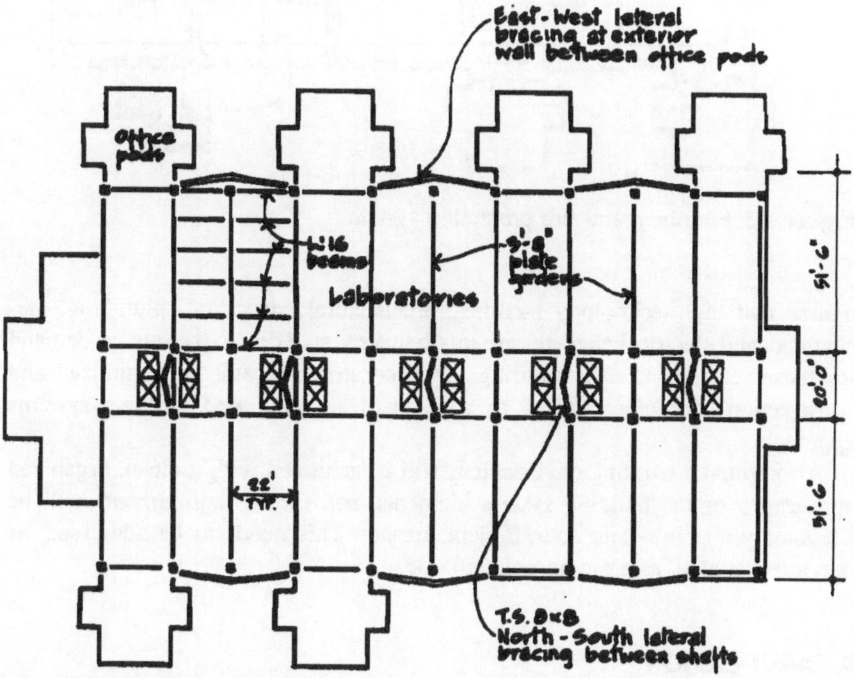

Figure 3.2. Structural system.

An integrated building systems approach, with or without interstitial space, is designed to provide for an increased level of adaptability (Figures 3.2 to 3.5). Its organizational plan provides dedicated zones within which all constructed elements are defined vertically and horizontally to control their present location and to insure their future availability when change is required. The system must rec-

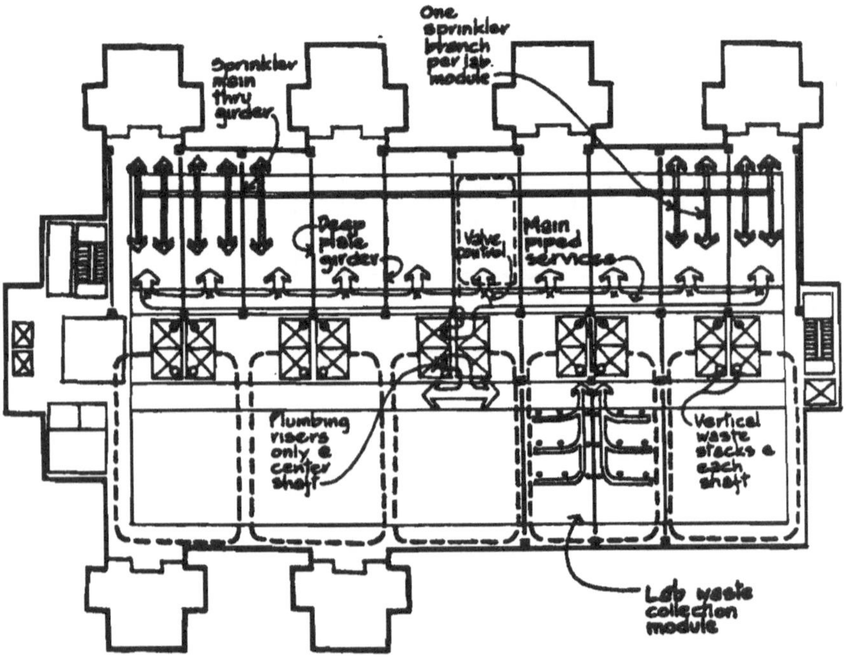

Figure 3.3. Plumbing and fire protection system.

ognize that high-technology facilities, architectural, structural, plumbing, mechanical, and electrical elements are totally interrelated due to the intense demand for these services. Conflicts during initial construction will be minimized and construction schedules reduced as a result of an integrated building systems approach.

An improved working environment will be achieved with a clean, organized appearance of the building system elements, and future requirements will be adequately met in a safe and efficient manner. This needs to be addressed, as previously stated, as a programmatic issue.

4. CIRCULATION

Circulation within and among laboratories is another component of safety often undervalued. Life safety within a laboratory can be measured in degrees, but, taken in the extreme, there are those occasions when accidents occur and a clear and orderly egress is both paramount to the safety of the users and mandatory in the context of good planning.

Our recommendation is to provide two means of egress in opposite directions from any point within the laboratory.

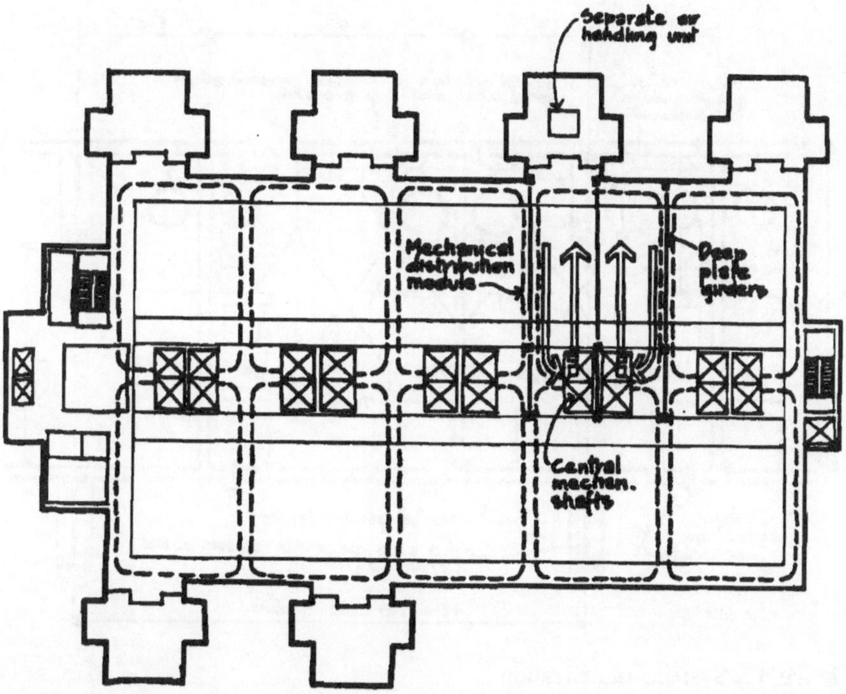

Figure 3.4. Mechanical distribution system.

In an optimal sense this means that the laboratory space is flanked on each end by a dedicated corridor system (Figure 3.6). The nature of the construction of these corridors is governed in different ways by different jurisdictions, but for the most part only one corridor may be required to be of a fire-rated construction value.

5. SERVICE CORRIDOR

The service corridor can be designed to be adequate in width and serviceable in detail to accommodate the movement of large pieces of equipment, chemicals, and support materials.

The manipulation of a new 8-ft fume hood into an existing laboratory without demolition of existing partitions has been one of the classic problems of facility design. With this added corridor width (assuming appropriate door width and height) this now becomes a more feasible undertaking.

The finish treatment of this corridor, walls and flooring, can also be made to be responsive to the abuse to which this area of a building is typically subjected. The virtual armor plating of surfaces can significantly diminish the normal deg-

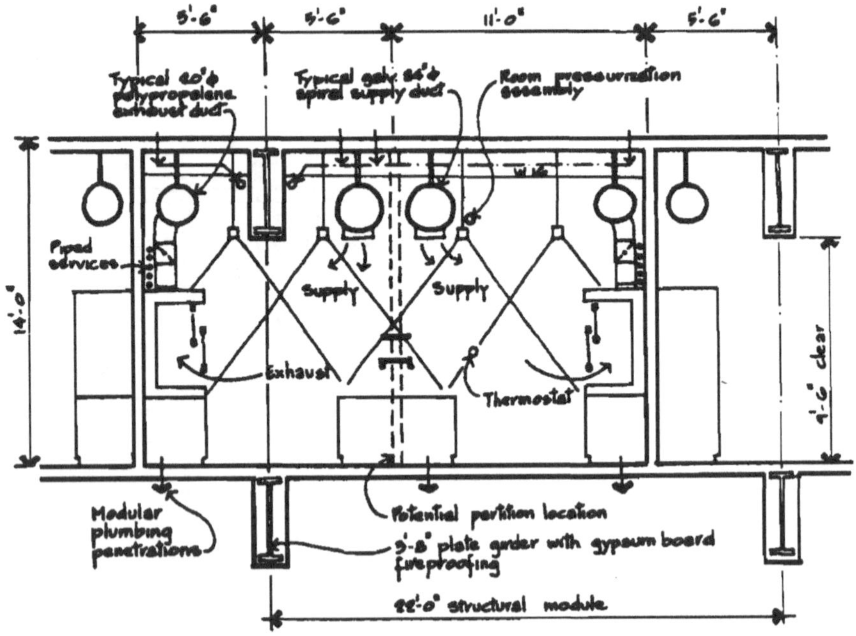

Figure 3.5. Systems organization.

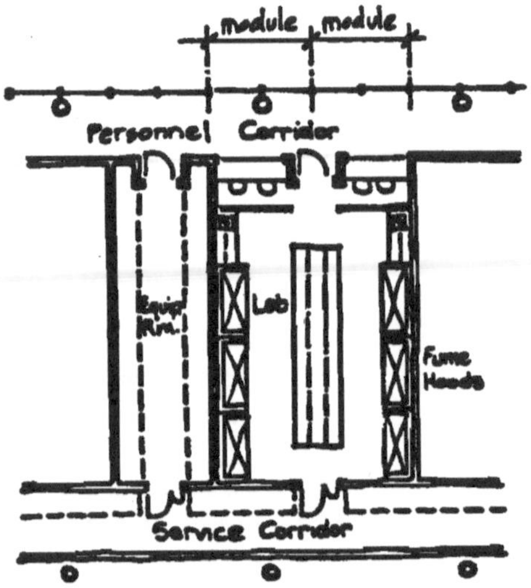

Figure 3.6.

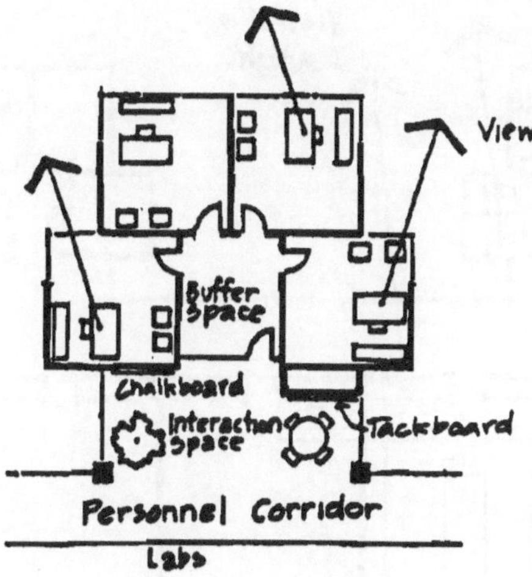

Figure 3.7.

radation of the integrity of surfaces, and because it is a transition zone within the building, it does not jeopardize the humane aspect of normally occupied space.

Assuming the corridor is not part of the fire-rated existing system for the building, its added width can also be utilized as assignable space for equipment, bottled gases, and vented storage cabinets. These items can be parked adjacent to, but outside of, the lab which they may serve. This further aids in the diminishing of noise and vibration within the lab, removing one more element of the fatigue factor for work undertaken within that research component. It places an appropriate storage capacity for chemicals in convenient proximity to the lab, precluding the need for storage of same in the lab within the hood or worse. It also provides space for more bench length within the lab, further minimizing clutter on what is always precious real estate.

6. PERSONNEL CORRIDOR

Companion to the service corridor in the two-corridor system is the personnel corridor. Environmentally, this is the zone dedicated solely to the movement of people and, as such, is detailed to be responsive to more humane needs. The potential for carpeted floors, special lighting, and areas of interaction are the highlights of this type of space. To succeed, however, takes a commitment to the proper operation, precluding the movement of chemicals and equipment within this space. It is truly and completely a zone dedicated to people (Figure 3.7).

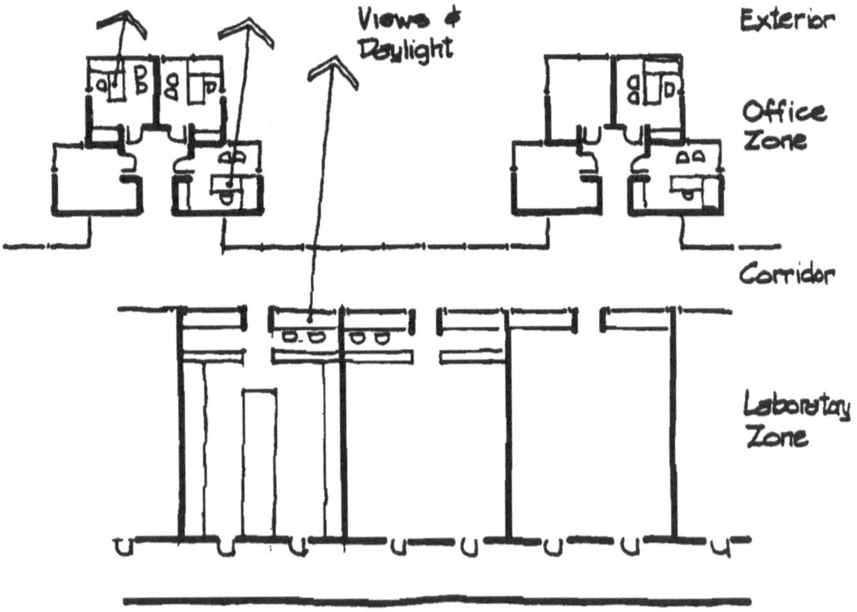

Figure 3.8. Exterior awareness/views and daylight.

7. DAYLIGHTING AND VIEWS

Within the laboratory, beyond the need to provide for the technical elements which respond to the functional requirements of the users, there is an equally compelling and often overlooked requirement for a humane environment. Fatigue is often the key ingredient in provoking accidents within the lab. Much of this fatigue can be contributed to the hazards of sensory deprivation, a monotony attributable to some manmade environments where users are encapsulated in sterile spaces without visual relief from the tedium of their endeavors.

Certainly, appropriate levels of artificial lighting — strategically placed and functioning within an acceptable temperature range providing a full color spectrum — can go a substantial way in reducing the problem.

There are, however, other considerations which must be evaluated when designing a research facility.

We strongly recommend that natural daylighting be given the highest consideration as a design tool for further alleviating this problem (Figure 3.8).

Research has been undertaken which verifies that individuals confined to bland *environments will, over a reasonably short period of time, exhibit difficulties in* concentration, periodic feelings of anxiety, a loss of ability to judge time, and some distortion of reality. There are those who conclude that normal consciousness, perception, and thought can be maintained only in a constantly changing environment. Where there is no change, a state of sensory deprivation occurs and the capacity of individuals to concentrate deteriorates, attention fluctuates and lapses, and normal perception diminishes and fades.

What better way to stimulate change than to provide for a view to the outside, maintaining contact with an environment that, by nature, is ever changing while borrowing a portion of the energy required for lighting.

To be successful, a research facility requires a balance of areas which need to function in a dedicated fashion as zones of refuge and zones of confluence.

8. ZONES OF CONFLUENCE

Zones of confluence are spaces that promote interaction and the exchange of ideas. Such zones reinforce a team concept in research by providing both planned and spontaneous interaction. Clearly, spaces that accommodate planned activities such as conference rooms, libraries, and meeting areas are crucial to the sharing of ideas and resources; but a lively synergy can be fostered through the design of connecting spaces that encourage unplanned encounters with researchers from other disciplines. Such spaces include open stairwells that visually link two floors and their activities, alcoves, or anterooms within corridors, properly furnished, that provide for informal interchange and more public areas such as lobbies, cafeterias, or outdoor patios that promote social communication.

9. ZONES OF REFUGE

Zones of refuge provide for privacy, contemplation, and concentration on individual tasks. Successful areas can be achieved by careful planning; providing buffering or acoustical separation or isolating space along a cul-de-sac to preclude audio or visual disruption.

10. SUMMARY

Safety in a laboratory environment is more than designing a facility to meet minimum life-safety codes. It also includes an understanding of the very nature of research and a functional and aesthetic response to the needs of the investigators themselves. This means architecture that cultivates a sense of community and enhances the users' image of themselves as professionals and contributors to society.

Laboratories conceived with safety in mind are those that can adapt to program change. Effective laboratories are designed on a modular basis with all systems, mechanical, electrical, structural, and architectural rationally planned for ease of modification. Thus, safety is provided for in the initial stages of a facility's life and maintained in future phases.

Circulation is a key design consideration. The use of separate corridors, one for service and equipment and one assigned as a personnel corridor, offers advantages in terms of supply and maintenance as well as establishing zones of interaction. This approach also offers two means of egress from the laboratory and thus provides additional safety advantages.

In the environmental sense, safe laboratory design means the creation of space that is stimulating; design should include natural lighting and views to the exterior to preclude fatigue and thus promote safety.

Safety, then, is an essential component of laboratory design and contributes to the overall success of a research facility as a place of inspiration, investigation, and problem solving.

CHAPTER 4

Safe Laboratory Design: A Safety Professional's Perspective

Douglas B. Walters

1. INTRODUCTION

The scientific community has shown an increased concern in safe laboratory design during the last 10 years.[1-8] This observation can, in part, be attributed to an increased awareness over the safe use and handling of hazardous chemicals, as well as new regulations and an upsurge of public concern regarding chemical safety. As a result, more than ever before, when new laboratories are built or refurbished, safety requirements are included in the initial planning along with the work requirements of the laboratory. However, we are still a long way from being able to state that safety concerns are always routinely included in the initial planning stages of laboratory design. Good and safe laboratory design still remains an often elusive aspiration.

Historically, when chemistry was in its infancy, little regard was given to laboratory design, let alone the safety of the workers or the environment. Slowly, conditions changed and buildings were designed specifically for use as laboratories. However, the problems of worker and environmental safety persisted. Today, we are finally beginning to realize that proper laboratory design necessitates including safety initially, as well as throughout the entire design and construction stages. Table 4.1 lists those fields which necessitate close interdisciplinary cooperation and interaction with health and safety.

Consistent with all of these disciplines must be the common goal of safe

Table 4.1 Disciplines Requiring Close Interaction with Health and Safety

Architecture
Human factor engineering
Electrical engineering (e.g., special studies such as inhalation toxicology)
Industrial hygiene
Safety specialties (e.g., fire, etc.)
Specific scientific disciplines (e.g., chemistry, biology, veterinary medicine, etc.)
Toxicology
Regulatory expertise
Legal expertise
Management

laboratory design. In that respect, we are all co-workers and need to be considered equally. The obvious significance of certain disciplines, such as architecture and ventilation engineering, in laboratory design is understood. However, the inclusion of other areas may require a brief explanation.

The most overlooked field is human factors or ergonomic engineering, which is the study of the individual in the work place.[9] This approach permits the worker to perform operations with a minimum of stress and a maximum of efficiency. It should be remembered that, not only are happy workers more productive, but they are also safer. Such considerations as proper lighting, comfortable as well as anatomically supportive chairs, knee wells to provide leg room, and appropriate work station design always need to be made a part of design planning. A large measure of the health effects controversy over the use of video display terminals (VDTs) has been attributed to improper ergonomic input in the design of these work stations.[10]

Although the importance of consulting with specific laboratory users for their requirements is obvious, this is not done; or it is not done in a complete fashion. For example, the head of a chemistry group may be asked for general input, but the lone synthetic chemist using vacuum racks or the natural products chemist requiring large extraction or distillation apparatus may be overlooked. Similarly, animal care specialists and veterinarians must be solicited for their particular requirements, not only for their studies, but also for humane animal treatment considerations.

Users of special equipment such as mass spectrometers, NMRs, lasers, etc., must provide specific information on weight, size, electricity, and any other precise requirements. Inhalation toxicologists frequently have unique spacial, as well as ventilation and electrical needs. Advice in the beginning from electrical engineers familiar with these specialized requirements pays for itself by preventing future problems. Similarly, toxicologists always have special water, lighting, ventilation, and structural requirements.

Anyone who has ever been engaged in the problems of getting an incinerator

Table 4.2 Areas Usually Considered as Nonlaboratory Areas

Archives
Animal support areas
Conference rooms
Dry ice storage
Eating/break areas
Elevators
Equipment storage
First aid office
Gas cylinder storage
Janitorial supplies
Library
Maintenance engineering
Nonchemical waste areas
Offices
Photography/dark rooms
Physical plant
Receiving
Security control areas
Training rooms
Copy rooms

certified, writing an environment impact statement, or the difficulties of modifying local covenants or obtaining special building permits, understands the importance of legal advice. The same applies to the need for expert insight on regulatory matters and sound management input.

The technical areas which frequently present the most problems during the design of safe laboratories are

- Ventilation — in general
- Hoods and vented work stations
- Chemical handling
- Chemical storage
- Waste disposal and storage

Other areas often neglected in the initial planning phase include janitorial closets, chemical and waste storage areas, shops, and break rooms. Design standards for first aid stations, medical examination rooms, and nurses stations are exceptionally rigid for certification. Elevators are not usually considered from the standpoint of transporting chemicals which may become hazardous if spilled in such confined areas. Similarly, volatile chemicals and cryogenics, such as dry ice, may be particularly hazardous in the elevator stalls between floors for any length of time.[11] A partial list of often neglected areas usually considered as nonlaboratory areas is shown in Table 4.2.[12]

2. HEALTH AND SAFETY CONCERNS

Almost without exception, every laboratory is a unique entity; each possessing its own particular design requirements. Just as it is important to obtain user input to provide valuable insight to the architect and engineer, it is also imperative that qualified, professional safety advice be sought.

The discipline of health and safety has four objectives:

- To protect the workers
- To safeguard the environment
- To comply with regulations
- To assist in attaining study goals

Worker and environmental protection have always constituted major portions of the responsibilities for the safety offices of any organization. From a laboratory standpoint, this means guarding against excessive exposure to chemicals, their metabolites, and their degradation products. In addition, personnel and the environment must be protected from contamination by other hazardous agents such as biohazardous materials, ionizing, and nonionizing radiation. Similarly, as a result of today's rapidly changing requirements and increasing complex regulations, compliance demands with the many local, state, and federal regulations have grown exponentially. This is particularly true when various guidelines and recommendations, which also require response, are considered.[13-16]

One of the most underutilized responsibilities of safety professionals is their function in helping to achieve the goals of the study. Too often, the safety staff are considered as adversaries or policemen, rather than as co-workers. This misconception occurs because of their responsibilities for enforcement of policies, rules, and regulations. However, safety professionals are unique because they are usually familiar, in detail, with all aspects of the varied work requirements within a particular organization, as well as having an intimate knowledge of the facilities. This information is a particular asset when the safety professional serves as the coordinator when new facilities are constructed or refurbished. Their function, in this capacity, is to serve as intermediaries to facilitate communication between the architects and engineers and the scientific staff.

3. DESIGN PHASE

Proper laboratory design is an iterative process which occurs in stages. It is imperative that health and safety be included in each stage. The five stages necessary for effective laboratory development are

- Definition
- Interpretation
- Design
- Construction
- Occupancy

Definition Phase

In the definition stage, all personnel who will be using or occupying the facility are asked for input. This input must be a realistic and truthful estimate of their requirements and must consider the four components of design, namely, space, mechanical, structural, and equipment.[17]

It should be stressed to the laboratory users that the more completely their requirements are expressed to the designers, the easier it will be to understand and to satisfy their needs. The importance of doing their homework saves time and avoids costly misunderstandings. Design considerations by the users must include work classification, operations performed, materials and amounts handled, and staff size. Safety is an inherent concern for each of these categories and must be taken into consideration in the beginning of the design phase.

Interpretation Phase

After the design requirements are communicated to the architects and engineers, preliminary plans and drawings are prepared. These ideas are then presented to the users of the laboratory again for review and additional input. It is at this interpretative stage that the value of the safety professional takes on importance. Experience has indicated the necessity at this point of identifying an individual who has both in-depth familiarity with the work for which the facility is being designed, as well as knowledge, experience, and interest in design and engineering.

The purpose of this individual is to serve as a liaison between the scientific staff and the designers. Frequently, it is necessary to resolve differences of opinion or to insist that certain costly equipment be maintained. These differences occur not only between the staff members and the architects, but also between the staff members themselves, especially when cost is an issue. It is at this point that an individual is needed who can understand the concerns of both sides, and who possesses both technical expertise and experience.

Frequently, it is the safety professional who meets these requirements and serves to negotiate and reach agreements over specific requirements and modifications. This phase is called the interpretative phase because it may require development of more than one initial set of preliminary drawings before a final design plan is agreed upon. This task is the only step which can effectively translate scientific specifications into pragmatic reality. The effort spent is always recovered later by minimizing timely and costly renovations which may be necessary after construction is completed.

Design Phase

The better and more complete the final design, the less likely that extensive changes and modifications will be necessary. While changes and oversights

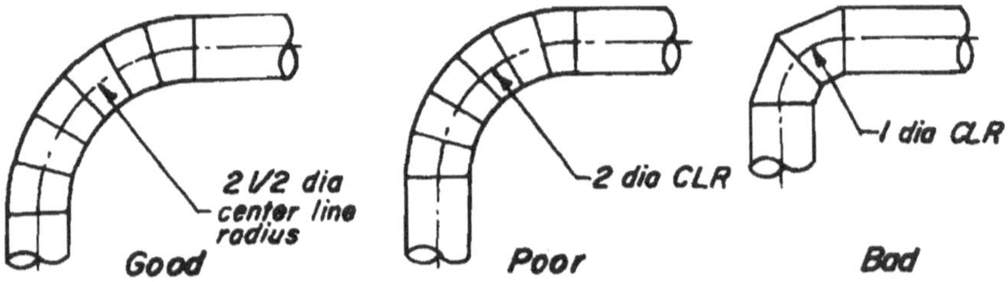

Figure 4.1. Proper and improper duct elbow bends.

always occur, it should be remembered that the earlier these changes are identified in the planning stages the less costly, inconvenient, and time consuming they are to effect. Once the final plan is decided on, the process moves into the next phase.

Construction Phase

The importance of daily monitoring by the safety professional-coordinator during the construction phase cannot be overemphasized. It is common that the agreed upon plans of users and designers are not properly translated and carried out by the craftsmen doing the construction. For example, duct work cannot always be implemented as designed due to unforeseen interferences. In other instances, well intended craftsmen do not always construct the actual duct work to the specifications of the engineers. Other problems commonly encountered include: substitution of unsuitable materials for those specified, improper placement of equipment or air handling filter banks which make maintenance difficult or impossible. Figures 4.1 to 4.3 show examples of properly and improperly designed duct work from an industrial hygiene standpoint. [15]

If these basic principles are not instituted, the resultant ventilation system will be inefficient, much more expensive to operate, and worst of all, may not accomplish what it is supposed to do, i.e., capture and remove hazardous agents and protect the worker and the environment. It is usually only a safety professional who has sufficient knowledge to make these decisions.

Occupancy Phase

Once construction is completed, a detailed inspection is necessary prior to occupancy of the building. In addition to typical inspections for compliance with building codes and various local, state, and federal regulations, any other changes deemed necessary should be done at this time, if possible. It is much easier, quicker, and cheaper to make changes before a facility is occupied than afterwards.

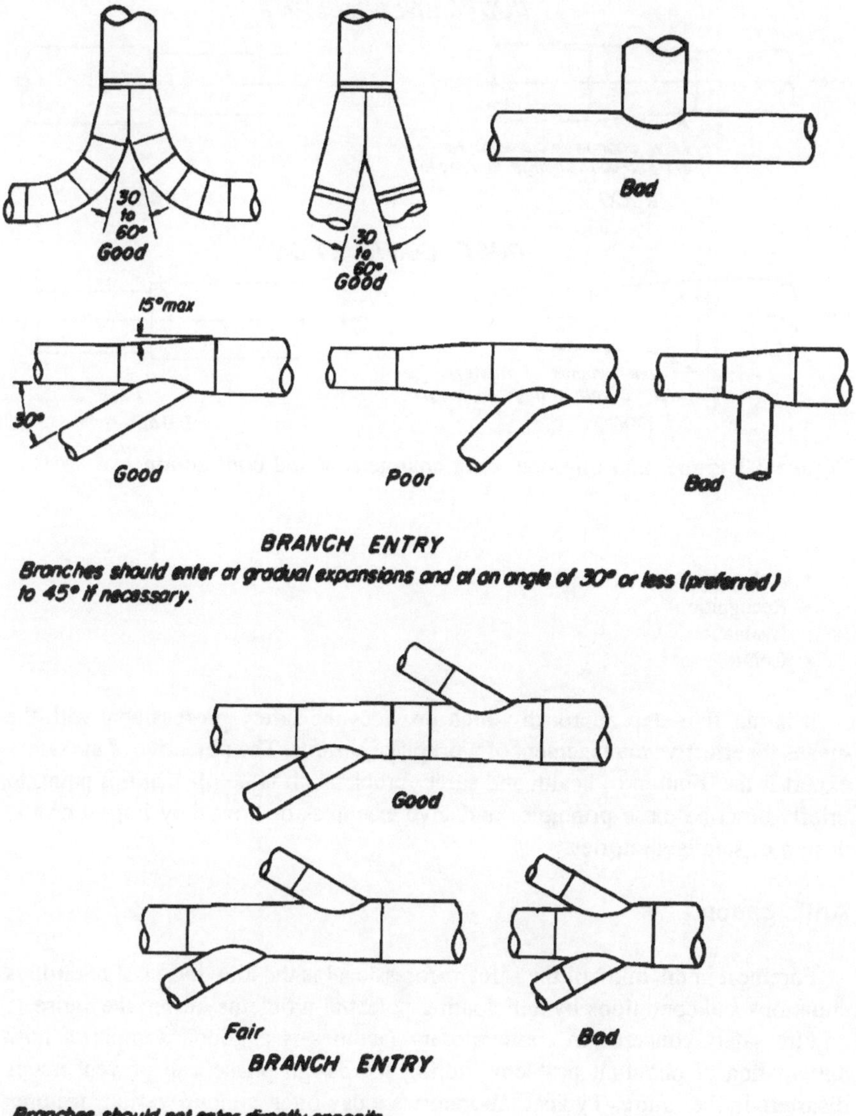

Figure 4.2. Proper and improper duct branch entries.

4. THE PRINCIPLES OF INDUSTRIAL HYGIENE

The basis for the control of safety related laboratory problems is the applicaton of the fundamental principles of industrial hygiene, namely:

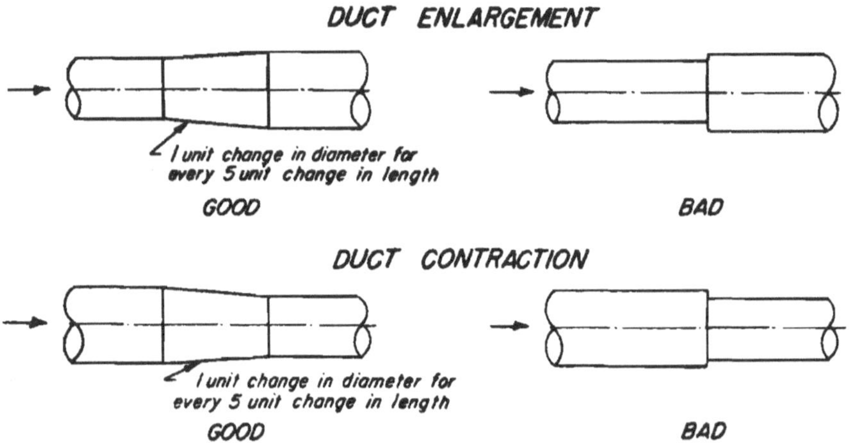

Figure 4.3. Proper and improper duct enlargement and contraction.

- Anticipation
- Recognition
- Evaluation
- Control

It is this four-step approach which provides the safety professional with the means for effective management of workplace hazards. The objective of the safety expert is the solution of health and safety problems. It is helpful, at this point, to briefly describe these principles and give examples of how they impact on the design of safe laboratories.

Anticipation

Foremost in the mind of the safety professional is the alleviation of hazardous situations and conditions by anticipating potential problems before they arise.

Fire safety concerns in contemporary facilities is a good example of how anticipation of potential problems during the design phase can prevent major disasters in the future. Typical laboratories today often contain various features such as: limited access areas, barrier facilities of the type found in toxicology laboratories, interstitial spaces, drop ceilings, and false walls. All of these features increase the hazard, risk, and damage caused in the event of a fire. The basic strategy of fire fighting has changed little in recent times and can be separated into the following seven factors applicable to all fire situations: access; rescue; locating; confining and extinguishing; exposure; ventilation; salvage and overhaul.[18]

In planning and designing laboratory facilities, these factors must be integrated into the design process. The design team, with the help of the safety professional, must anticipate emergency fire conditions by considering the type and amount of

Figure 4.4. Examples of improper weather cap design.

fuel, its location, possible ignition sources, routes of fire and smoke movement, detection and alarms, fire suppression systems, communication, evacuation routes, access for firefighters, structural stability, confinement of hazardous materials, as well as many other factors.[18] This can only be done by fully qualifed and experienced safety professionals.

Recognition

Recognition of potential problem areas or situations is the second concern of the industrial hygienist. In this case an intimate knowledge of the work which is planned for the laboratory or facility is invaluable. Similarly, having a wide range of experience and interdisciplinary training is very useful. In a laboratory operation, not only is it important to know the precise chemicals used and their physical properties, but, knowledge of the technology to control exposure is also necessary. In additon, once a problem is identified it is necessary to formulate suitable alternatives to alleviate the condition.

Figure 4.4 shows an example of a laboratory rooftop where hazardous chemicals were being used. The stack in the foreground is commonly called a china cap or coolie cover and is used as a weather guard to prevent rain from returning into the ventilation system. A similar device, called a mushroom cap is shown immediately behind the china cap. In both these instances, the weather covers were used over ventilation ducts which exhausted effluent from hoods and laboratory areas where hazardous and volatile chemicals were used. Both types of weather caps are expressly not recommended,[15] because they force the effluent air back down onto the roof top and can facilitate the reentrainment of the contaminants back into the facility.

Evaluation

Evaluation is the process whereby surveys or experimental determinations are used to evaluate the effectiveness of existing conditions or equipment. Once again, detailed knowledge of the materials used, their properties, and the requirements of the job are necessary. Figure 4.5 is a diagram of a Plexiglas® cover designed to be placed over an analytical balance and vented to the outside.[19] The balance was used for weighing highly hazardous materials. In designing the balance cover, input was sought from analytical chemists, technicians, ventilation engineers, human factor engineers, and industrial hygienists. The result was the prototype in Figure 4.6.

Simply designing the device was not sufficient, however. The task still remained to establish that it could be used without restricting the users ease of performing the weighing and that the user could obtain the desired accuracy without a corresponding increase in the time required to perform the task. In addition, it was necessary to determine the minimum airflow needed to ensure the protection of the worker and the environment. Finally, air sampling was performed using tracer compounds to ensure maximum containment. After sampling and analysis, as well as obtaining input from the users, improvements were made and a final design was produced. This exercise illustrates both the significance of evaluation and the importance of interdisciplinary teamwork.

Control

Once a problem is identified, the importance of controlling the problem becomes paramount. Figure 4.7 is an example of a situation where the effluent from a high hazard laboratory was exhausted from the building immediately adjacent to the air intake for the building.

The solution to such problems is usually relatively straightforward, depending on the nature of the effluent. Figure 4.8 shows an exhaust stack in a similar type of laboratory where a bank of filters was used to trap and contain airborne contaminants before they reached the stack. In this instance, HEPA filters, for control of particulate materials, were employed in combination with charcoal filters to control volatile organics. As an added precaution, an extra set of charcoal filters were placed behind the first set to ensure that if breakthrough did occur due to a leak or to overloading, the second set of charcoal filters would trap the contaminants.

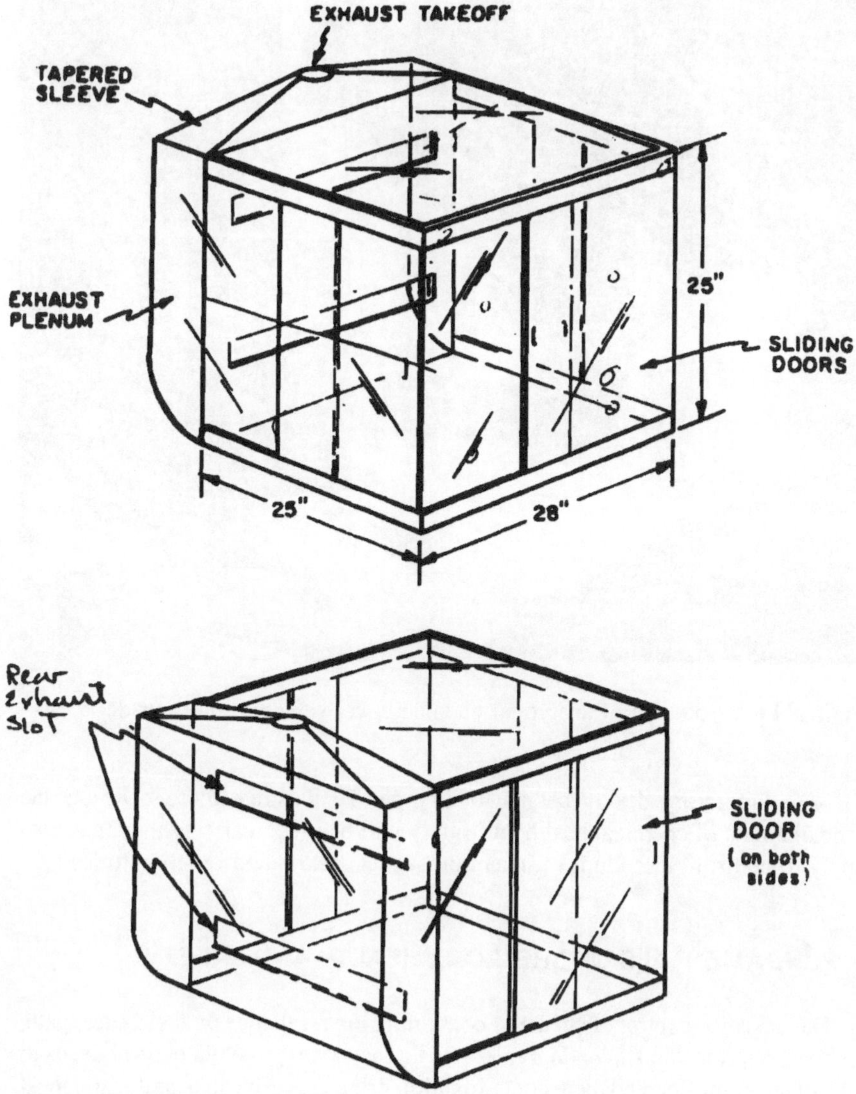

Figure 4.5. Design for a Plexiglas® vented balance cover.

The presence of the second set of filters demonstrates another important rule of safe laboratory design. Namely, the importance of redundancy or backup. Another look at Figure 4.8 indicated an entire parallel system of filters and blowers for the same effluent exhaust system. This was done so that when it became necessary to change out the filters, or if the blower developed a problem, the adjacent parallel unit could take over completely. Hence, another set of rules

Figure 4.6. Prototype of analytical balance cover vented to the outside.

for safe laboratory design begins to emerge. To illustrate these concepts the fundamentals of chemical health and safety will be used, realizing they may also apply to control other kinds of hazardous agents also found in laboratories.

5. FUNDAMENTALS OF CHEMICAL HEALTH AND SAFETY

The basis for control of chemcial contamination is reliance on the source, path, receiver relationship shown in Table 4.3. To the extent possible, control needs to take place at the source where contamination originates. This is usually, and most practically done, by the application of engineering controls. Figure 4.9 is an example of a glove box used to control the spread of contamination to the surrounding area. The basic principle in this instance, is to simply enclose the source of contamination.

The basic sequence of the design concepts is to

- Understand the properties of the hazard controlled
- Design to control the hazards
- Design for the specific task(s)
- Design for efficiency

Figure 4.7. Examples of exhaust reentrainment.

In other cases, it is not always so simple to effect control of contamination at the source. As a result, some measure of compromise may be necessary. This can be seen in Figure 4.10.[20] This is a work station used for necropsy and tissue trimming of rodents in toxicology laboratories. The hazards present are volatile reagents and test chemicals to which the animals may have been exposed.

While it is theoretically possible to totally enclose the entire work area, the

Figure 4.8. Filter banks for control of effluent exhaust.

nature of the hazard, the high cost of enclosing, and the extreme loss of worker efficiency and work quality make this an unacceptable option. Instead, what was developed was a reliance on a combination of controls. Employment of engineering controls at the source, to the extent possible (i.e., enclosing and venting the area), should be coupled with reliance on good operational practice along the path of contamination. Operational practices means ensuring that equipment is used in accordance with good industrial hygiene techniques. These techniques include ensuring that all work in the ventilated area is performed in the center, and at least 6 in. back from the front edge, and that no contaminated materials or waste is placed outside the area in open containers (e.g., contaminated disposable pipets and paper wipes).

It should be explained at this point what is meant by the path of contamination. This is simply the path the contamination takes from the source of origin in order to eventually reach the endpoint of the system, or the receiver. In the case of worker exposure, the receiver is the worker.

Where the protection of the worker is still in question, even after application of sound engineering controls and operational work practices, it is necessary to institute the use of personal protective clothing and equipment. It is important to stress, however, that in no instance should reliance on worker protection rest solely on the use of personal protection. There are two reasons to avoid this serious error from an industrial hygiene viewpoint. First, no item of personal

Table 4.3 Health and Safety Concepts and Controls

		Concept		
Source	→	Path	→	Receiver
		Control techniques used		
Engineering control	→	Operational practice	→	Personal protection

Figure 4.9. Containment at the source of contamination.

protective equipment or clothing ever offers 100% protection. Second, in the event of equipment breakdown, breakthrough, or malfunction, there is no back up; in other words, there is no redundancy.

In addition to designing for the chemicals to be used and the work which is to be performed, the objective of chemical health and safety is to **MAXIMIZE CONTAINMENT AND MINIMIZE CONTAMINATION.** Remembering that, **REDUNDANCY IS THE KEY.**

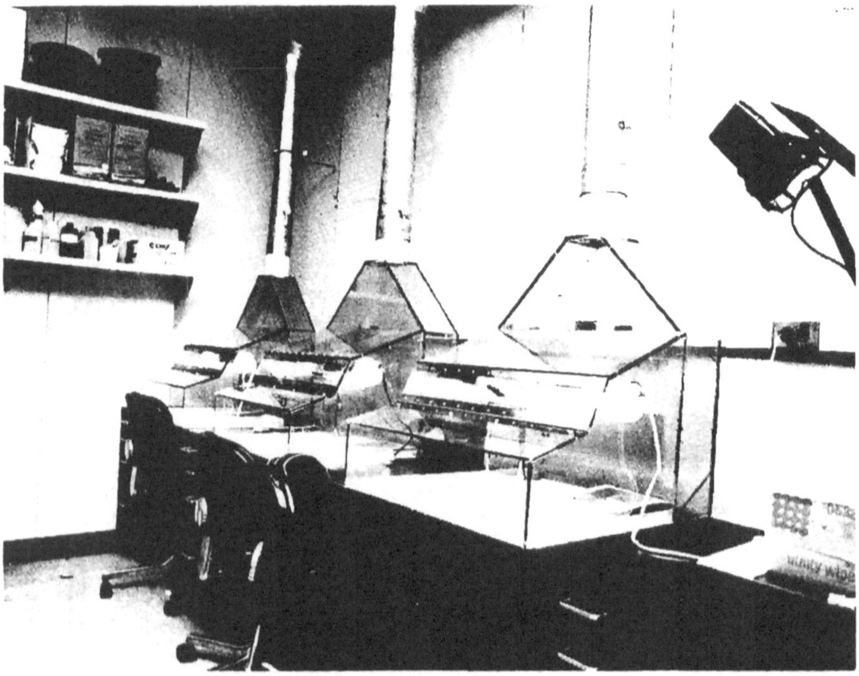

Figure 4.10. Control along the path of contamination.

6. CONCLUSION

Safe laboratory design is the practical application of the principles of industrial hygiene and chemical containment. Complete knowledge of the properties of the materials to be used in the laboratory and precise information on how and where they will be used is necessary. Today's complex environment demands close interdisciplinary cooperation of many different specialties all working together for the achievement of the common goal — a safe work place and a safe environment. There are few other instances in which so many basic concepts from so many different professions can be used so effectively.

References

1. Walters, D. B., Ed., *Safe Handling of Chemical Carcinogens, Mutagens, Teratogens and Highly Toxic Substances* (Ann Arbor, MI: Ann Arbor Science, 1979) Vols. 1 & 2.
2. Everett, K. and Hughes, D., *A Guide to Laboratory Design*, London: Butterworth, 1981.
3. Walters, D. B., Ed. *Health and Safety for Toxicity Testing*, (Boston, MA: Butterworth, 1984).

4. Teteris, A. S., Ed. *Toxicology Laboratory Design and Management for the 80's and Beyond* (Basel: Karger, 1984).
5. Lees, R. and A. F. Smith, Eds., *Design, Construction and Refurbishment of Laboratories* (West Sussex, England: Ellis Horwood Ltd., 1984).
6. DiBerardinis, L. J., J. Baum, M.W. First, G.T. Gatwood, E. Groden, and A. K. Seth, *Guidelines for Laboratory Design: Health and Safety Considerations* (New York: John Wiley & Sons, 1987).
7. Wadden, R. A. and Scheff, P. A. *Engineering Design for the Control of Workplace Hazards*, (New York: McGraw-Hill, 1987).
8. Scott, R. A. and L. J. Doemeny, Eds., *Design Considerations for Toxic and Explosive Facilities*, (Washington, D. C.: ACS Symposium Series, 1987).
9. Phelan, E. J., C. M. Snyder and D. B. Walters, "Human Factors Considerations in the Handling of Toxic Chemicals," in *Health and Safety for Toxicity Testing*, (Boston, MA: Butterworth Publ., 1984), D. B. Walters and C. W. Jameson, Eds., pp. 121-133.
10. *American National Standards for Human Factors Engineering of Visual Display Terminal Workstations*, ANSI HPS 100-1988.
11. Bretherick, Leslie, personal communication, 1988.
12. Lees, R. and A. F. Smith, Eds., *Design, Construction and Refurbishment of Laboratories* (West Sussex, England: Ellis Horwood Ltd., 1984, pp. 23-26.
13. *National Institutes of Health (NIH) Guidelines for the Laboratory Use of Chemical Carcinogens*, DHHS NIH-81-2385.
14. *Guidelines for Research Involving Recombinant DNA Molecules*, National Institutes of Health (1978).
15. *Industrial Ventilation, A Manual of Recommended Practice*, 20th ed. (Lansing, MI: American Conference of Governmental Industrial Hygienists, 1989).
16. *Prudent Practices for Handling Hazardous Chemicals in Laboratories*, (Washington, D.C.: National Academy of Sciences Press, 1981).
17. Harless, M. M., "Components in the Design of a Hazardous Chemical Handling Facility," in *Health and Safety for Toxicity Testing*, D. B. Walters and C. W. Jameson, Eds., (Boston, MA:, Butterworth Publ., 1984), pp. 45-71.
18. Nemchin, R. G., "Barrier Laboratory Facilities: A Fire Control Manager's Perspective," in *Health and Safety for Toxicity Testing*, D. B. Walters and C. W. Jameson, Eds., (Boston, MA: Butterworth Publ, 1984). pp. 91-110.
19. Hoyle, E. R., T. Murray, C. M. Snyder, A. T. Prokopetz, and D. B. Walters, "Design and Evaluation of an Exhausted Enclosure for an Analytical Balance," submitted to *Appl. Indus. Hygien.*, (1988).
20. Snyder, C. M., R. Tuomanen, D. B. Walters, and A. T. Prokopetz, "Design and Evaluation of a Tissue Trimming Work Station," submitted to *Appl. Indus. Hygien.*, (1988).

Section II: Generic Issues Affecting Design of Safe Laboratories

CHAPTER 5

An Overview of Basic Life Safety Principles

Peter Ashbrook, Paul Restivo, and Stephen Szabo

1. INTRODUCTION

One of the facts of life is that accidents happen. The purpose of life safety planning is to be prepared for such accidents and to minimize the ensuing consequences. Certain kinds of accidents (e.g., fires) have occurred so frequently that various codes and regulations have been established to guide design of new or remodeled laboratories.

This chapter will give an overview of life safety principles to give the reader guidance in applying the various codes and regulations. For the purposes of this chapter, life safety considerations have been divided into the following categories: construction requirements, fire prevention, general emergency considerations, worker safety, design and equipment standards, and environmental considerations.

For those without experience using the life safety codes, several basic references provide a good overview of the concepts and requirements. These include the "Prudent Practices" books;[1,2] the National Fire Protection Association (NFPA) laboratory code;[3] and the Occupational Safety and Health Administration (OSHA) standards.[4]

Although architects and engineers are hired to design buildings in compliance with all applicable codes, it is important for the user to understand the rationale behind these codes. Codes almost always provide only minimally acceptable levels of safety. Because the user is the person most likely to understand the potential hazards in the laboratory, the user may be able to identify specific items in which meeting the codes will not provide adequate protection against hazards.

2. CONSTRUCTION CODES

There are a wide variety of construction/building codes adopted by local jurisdictions in the U.S. By far, the two codes most frequently adopted are the Building Officials and Code Administrators (BOCA) National Building Code[5] and the Uniform Building Code (UBC).[6] In addition to BOCA and UBC, many local jurisdictions have developed codes of their own. One such code is the Chicago Building Code. Some jurisdictions have essentially adopted hybrid codes which incorporate desired features of BOCA, UBC, and other codes. In situations where multiple codes have been adopted, it is prudent to design the facility based on the most restrictive requirements.

Construction/building codes have been developed to offer basic guidelines for architects/engineers in the design of buildings, and the codes cover a multitude of issues. For example, if one is designing a semiconductor research facility using highly toxic chemicals such as arsine and phosphine, the code requirements will be much more detailed than say the construction of a general wet chemistry lab. The semiconductor research facility and general wet chemistry lab are examples of "occupancy classifications" as detailed in the codes. Thus, the starting point in the design of a new facility should be determination of the occupancy classification based on the intended use of the new facility. The occupancy classification will be the "common denominator" upon which other code requirements are built.

After determining the occupancy classification of the new facility, the codes address such factors as the type of construction, allowable floor area, and height and number of stories. Relatively speaking, the more fire safe a building is (e.g., fully sprinklered, noncombustible construction materials) the larger the building may be. On the other hand, if one intends to build a stick building (combustible building materials without a fire protection system), it will be found that both the size of such a stick building and the intended uses of the building will be severely restricted.

If one is building a particularly hazardous facility (e.g., the semiconductor research facility), the codes may well provide additional requirements related to the geographic location of the facility in relation to other buildings or public ways. These additional requirements are intended to provide another layer of safety in the event of a catastrophic release of highly toxic chemicals which can be immediately hazardous to humans at low airborne concentrations.

Construction/building codes address basic means of egress requirements including numbers of exits, travel distances to exits, and construction requirements for means of egress components (fire rated walls and doors). These requirements will be discussed later in this chapter.

A final note about construction/building codes is that it is prudent to plan

Table 5.1 NFPA Codes Most Applicable to Design of Laboratories

NFPA 30	Flammable and combustible liquids code
NFPA 45	Standard on fire protection for laboratories using chemicals
NFPA 70	National electrical code
NFPA 90A	Standard for the installation of air conditioning and ventilating systems
NFPA 91	Standard for the installation of blower and exhaust systems for dust, stock and vapor removal or conveying
NFPA 101	Code for safety to life from fire in buildings and structures

ahead. If one wishes the facility to be constructed so that its future use may be flexible, one might wish to build in additional safety features (e.g., fully sprinklering the building and using noncombustible construction materials). The cost of retrofitting the facility later will be prohibitively more expensive as well as being very disruptive to building users. If the additional costs can be borne now, it may well represent the best dollar investment decision you ever make.

3. FIRE CODES

As was true with construction/building codes, it is important to determine which fire codes have been adopted by the local jurisdiction. This discussion will concentrate on National Fire Protection Association codes although other codes such as the Uniform Fire Code may be adopted by localities. The core code of NFPA is NFPA 101 (Code for Safety to Life From Fire In Buildings and Structures — "The Life Safety Code").[7] The Life Safety Code was developed "... to establish minimum requirements that will provide a reasonable degree of safety from fire in buildings and structures." (NFPA 101-1-2.1)

NFPA 101 details *minimum requirements* for such features as means of egress; fire protection including construction and compartmentation, smoke barriers, special hazard protection, and interior finish; fire detection and suppression; and details specific requirements as related to various occupancies (e.g., assembly, health care, educational, business, and industrial).

In addition to the general requirements outlined in the Life Safety Code, a number of other NFPA codes have direct applicability to the design of laboratories (see Table 5.1).

For example, NFPA 45 is the Standard on Fire Protection for Laboratories Using Chemicals. NFPA 45 places laboratories into three hazard classes according to the quantities of flammable and combustible liquids used. The hazard class is used to determine such requirements as construction materials, laboratory size, amounts of flammable liquids stored, and egress requirements.

The use of sprinklers allows for greater flexibility in the size of the laboratory and use of flammable materials within the lab. Sprinklers are highly effective in controlling fires and containing property damage. In most cases, water is a suitable extinguishing agent even when flammable liquids are used. The major caution in using water is to make sure that water will not come into contact with water reactive chemicals such as elemental sodium or potassium. Depending on the individual circumstances, the issue of water reactive chemicals can be addressed through adoption of certain work practices (requires training), through use of an alternative extinguishing agent (expensive), or not using sprinklers at all (less protection for laboratory and building).

According to NFPA 45, high hazard (Class A) labs must have a second exit if the lab is larger than 500 ft^2. With a few exceptions, lower hazard labs are not required to have a second exit unless they have 1000 ft^2. Other factors affecting exit requirements include whether a hood is located next to the exit (not a good idea for a number of reasons), the use of compressed gas cylinders, and the use of cryogenic containers.

NFPA 45 addresses a number of other issues related to laboratory design. Desk areas should be designed so that they do not encourage under-desk storage. Restraining techniques should be used for items stored above eye level. Easy access to utilities should be provided. Slip-resistant floor surfaces should be considered. There should be adequate provision for the storage of incompatible materials and the various types of special wastes.

Another example of fire codes is NFPA 30, which prescribes limits for the storage of flammable liquids. Unless very small quantities are used, flammable liquids should be stored in safety cabinets. The amount allowed to be stored depends on the flammability hazard of the liquids and the type of container. Laboratories planning to use flammable liquids should take care that adequate space is provided for storage cabinets. If particularly large quantities of flammable liquids are to be used, consideration should be given to designing a dedicated flammable storage room.

NFPA 30 does not require that flammable storage cabinets be vented for fire protection purposes. However, if the cabinet is vented for other reasons (desirable for nuisance odors and potential health hazards) it is to be vented to the outdoors in a way that does not compromise the cabinet's protection against fire. Also, NFPA 30 indicates that if the cabinet is not vented, the vent openings must be sealed with a metal bung. Local codes should be consulted in relation to the issue of venting of storage cabinets.

One of the most frequent causes of fires is electrical. Fires may be the result of over-use of electrical circuits or the use of temporary wiring (extension cords) rather than permanent wiring. Compliance with the National Electrical Code (NFPA 70) represents the best means of elimination of potential electrical fires and NFPA 70 should be the referenced code for the laboratory buildings.

4. WORKER SAFETY

The Occupational Safety and Health Administration (OSHA) has developed detailed standards addressing work practices, personal protective equipment, exposure limits, and facility construction requirements. This discussion will concentrate on the construction requirements.

The majority of OSHA's standards associated with facility construction requirements were adopted from National Consensus Standards, such as the Life Safety Code, which were in effect when OSHA was established in 1970 by the Federal Occupational Safety and Health Act. Many of the OSHA standards have remained unchanged since their initial adoption; thus, the OSHA standards may differ from the code which was the original foundation of their standard.

OSHA standards address such construction requirements as:

- Walking and working surfaces
- Means of egress requirements
- Ventilation requirements
- Engineering controls to reduce noise
- Design of flammable and combustible liquid storage areas
- Machine guarding
- Restroom — minimum numbers of toilet facilities
- Requirement for eye wash and safety showers

Because OSHA is primarily concerned with worker safety, many of its standards are performance oriented, and reference other codes and standards discussed elsewhere in this chapter.

OSHA has proposed a new performance-oriented standard entitled: "Health & Safety Standards; Occupational Exposures to Toxic Substances in Laboratories".[8] This proposed standard recognizes that OSHA's "General Industry" standards currently treat laboratories essentially the same as an industrial facility. The proposed standard attempts to rectify this by developing performance-oriented requirements for the development of a "Chemical Hygiene Plan" for each employer. The plans would provide a written program which details how the employer intends to protect his/her employees from toxic chemicals in the lab. An excellent resource for safety in laboratories is *Safety in Academic Chemistry Laboratories* published by the American Chemical Society.[9] The title is somewhat deceiving since the contents of the book would be relevant to any laboratory.

Another important consideration related to worker safety in a laboratory is the design of the lab's ventilation system. An appropriately designed ventilation system in conjunction with proper work procedures, can help to protect lab workers from hazardous levels of contaminants. The type of ventilation system will vary greatly depending upon the intended use of the lab. Examples would include fume hoods, glove boxes, clean rooms, and always adequate "make-up air." The most widely used and respected guide for the design of ventilation systems is *Industrial Ventilation: A Manual of Recommended Practice*[10] published by the American Conference of Governmental Industrial Hygienists. Ventilation issues are discussed extensively elsewhere in this book.

Table 5.2 Selected Standards Applicable to Laboratories[11]

Marking Physical Hazards, Safety Color Code for, ANSI Z53.1
Occupational and Educational Eye and Face Protection, Practice for, ANSI Z87.1
Drinking-Water Coolers, Safety Standard for, ANSI/UL 399
Fixed Ladders, Safety Requirements for, ANSI A14.3
Design Loads for Buildings and Other Structures, Minimum, ANSI A58.1
Making Buildings and Facilities Accessible to, and Usable by, Physically Handicapped People, Specifications for, ANSI A117.1
Demolition, Safety Requirements for, ANSI A10.6
Automatic Sprinklers for Fire Protection Service, Safety Standard for, ANSI/UL 199
Emergency Lighting Equipment, Safety Standard for, ANSI/UL 924
Eyewash and Shower Equipment, Emergency, ANSI Z358.1
Lasers, Safe Use of, ANSI Z136.1
Industrial Lighting, Practice for, ANSI/IES RP7
Office Lighting, Practice for, ANSI/IES RP1

5. DESIGN AND EQUIPMENT STANDARDS

Another category of standards which covers the design of equipment are American National Standards Institute (ANSI) Standards. A partial listing of ANSI Standards is outlined in Table 5.2.

For example, ANSI Standard Z358.1 for eyewashes and safety showers specifies that units be located within 25 feet or 15 seconds of travel time from where an accident may occur. For the most part, individuals responsible for the construction of a new laboratory may not have to be intimately familiar with all of these ANSI Standards. However, it is important to determine which of the ANSI Standards should be referenced for a particular project and to recognize that they have a bearing on the design of laboratories.

6. ENVIRONMENTAL REGULATIONS

Besides these building-oriented codes and standards, the person involved in upgrading existing laboratories or constructing new laboratories also needs to be aware of, and sensitive to, the various environmental regulations that can impact on laboratory operations. The standards that are mentioned here are primarily federal standards. The reader should be aware that many states and localities have established standards that are more rigorous than the federal standards.

Environmental standards typically will cover the areas of air, water, land use, and waste disposal. Air standards are found in the Clean Air Act and in Title III of the Superfund Amendments and Reauthorization Act (SARA). Provisions of SARA Title III are often referred to as the Community Right-to-Know Act.

Recent public concerns about air toxics may lead to requirements for control of fume hood emissions.

Standards protecting water quality may be found in the Clean Water Act and Underground Storage Tank Regulations. Also, both state and local governments may have regulations dealing with sanitary waste disposal and surface waters run-off control. Land use controls are often imposed by local planning boards or zoning agencies.

Hazardous waste control typically will be found in areas like the Resource Conservation and Recovery Act (RCRA), Toxic Substances Control Act (TSCA) (PCBs and asbestos), state health department regulations, and state department of natural resources regulations.

These environmental regulations will not necessarily specify what construction options are to be addressed, but the impact of the regulations may have a bearing on how changes are made or how new construction is done so that the least costly means of complying with existing and future environmental concerns is provided.

7. SUMMARY

As a last word of caution, do not depend on the architect or engineering firm to keep your project in compliance with codes and with good laboratory practice. The owner/user must take an active role in planning for code compliance. Remember that codes and standards usually provide a minimally acceptable level of safety and are often difficult to apply to unique situations. It is important to understand the underlying concerns behind the codes, so that in cases where the requirements are unclear, good decisions with respect to safety can be made.

References

1. *Prudent Practices for Handling Hazardous Chemicals in Laboratories,* (Washington, D.C.: National Academy Press, 1981).
2. *Prudent Practices for Disposal of Chemical From Laboratories,* (Washington, D.C.: National Academy Press, 1983).
3. NFPA 45, *Standard on Fire Protection for Laboratories Using Chemicals,* National Fire Protection Association, Batterymarch Park, Quincy, Massachusetts.
4. *Code of Federal Regulations,* Chapter 29, Part 1910, Most Recent Edition. OSHA has prepared a summary of these standards entitled, "General Industry Digest" (OSHA Publication 2201). One free copy of each can be obtained from any OSHA regional office.
5. The BOCA National Building Code (with its companion, BOCA National Mechanical Code). Building Officials and Code Administrators (BOCA), 4051 West Flossmore Road, Country Club Hills, Illinois 60477, (312) 799-2300.

6. The Uniform Building Code (with its companion Uniform Mechanical Code). International Conference of Building Officials, 5360 South Workman Mill Road, Whittier, California 90601, (213) 699-0541.
7. NFPA publications are available from: National Fire Protection Association, Batterymarch Park, Quincy, Massachusetts, (617) 770-3000.
8. *Federal Register*, July 24, 1986, Vol. 51, #142, p. 26661.
9. *Safety in Academic Chemistry Laboratories*, American Chemical Society, 1155 16th Street, N.W., Washington, DC 20036, most recent edition.
10. *Industrial Ventilation: A Manual of Recommended Practice*, published by the American Conference of Governmental Industrial Hygienists, 6500 Glenway, Building D-5, Cincinnati, Ohio 45221. (Request the most recent edition.)
11. ANSI Standards are available from: American National Standards Institute, 1430 Broadway, New York, New York 10018, (212) 354-3300.

CHAPTER 6

Basic Principles of Ventilation In Chemistry Laboratories

Gerhard W. Knutson

1. INTRODUCTION

Control of chemical exposure to chemists or technicians in the laboratory setting can be achieved by a dual approach of education and engineering controls, primarily ventilation. It should be noted that education was listed before engineering controls. It is not possible to provide adequate protection within the laboratory if the chemist or technician does not adequately understand the purpose and function of the engineering controls and work to optimize controls. Any design can be compromised to the extent that protection is no longer provided through operator misuse or non-use.

To provide adequate education, a basic understanding of ventilation systems, the major control method, must be achieved. Simply stated, ventilation is the purposeful movement of air. To understand the basics of ventilation systems, airflow must first be investigated.

2. AIRFLOW

From an engineering point of view, air is an incompressible fluid. Since air is not seen and is most often not felt (or at least is so familiar that we pay no attention to it), we tend to disregard air. It may be advantageous to visualize air by smoke visualization tests or by analogy with a familiar incompressible fluid, water. Of greatest significance is the importance of turbulence caused by obstacles within the fluid flow.

We are all familiar with the eddy currents caused by an obstacle, such as a rock, in a fast moving stream. As the water parts and flows around the rock, a highly turbulent area is formed downstream of the rock. Vortices are formed and the water frequently moves upstream towards the rock.

In a similiar manner, air flowing past a person will have negative depression downstream which will cause the formation of vortices and frequently backflow eddys. This is especially important as the air moves past a person in front of a laboratory hood or other exhaust ventilation system. When the vortices are formed near the work and can entrain contaminants generated within the hood and the backflow eddys draw these contaminants toward the chemist or technician at the hood, significant exposures can result.

3. VENTILATION SYSTEMS

A typical laboratory has several types of ventilation. To understand the interplay between these systems, each must be examined separately.

Local Exhaust

Local exhaust provides control of contaminants by drawing air from the laboratory and discharging it outside, sometimes after passing through an air cleaning device. Greater efficiency in control is achieved when the exhaust device surrounds or encloses the source. Since gases and particles small enough to be breathable essentially follow the airflow, control of the air provides adequate control of the contaminant.

Devices such as bench slots, canopy hoods, backdraft benches, elephant trunks, and especially laboratory fume hoods can provide local exhaust of contaminants.

General Exhaust Ventilation

In some instances, the source of contaminant is not localized and therefore not amenable to local exhaust ventilation. In these cases, large quantities of air need to be exhausted from the room to provide adequate ventilation and ensure the airborne contaminant level is relatively low.

Comfort Ventilation

In many laboratories, there are significant heat sources which can cause an elevation of the workroom temperature. Exhaust ventilation can be provided to control heat and humidity to provide comfort to the operators.

Supply Ventilation

Whenever air is exhausted from a laboratory, an equal quantity of air must be supplied. If the supply is not done mechanically then supply will occur through infiltration, backdrafting of inactive hoods, or other undesirable methods. Consequently, a supply system is required to ensure creature comfort and proper operating of the exhaust ventilation systems.

In most heating, ventilating, and air conditioning (HVAC) work, the systems are designed to cause significant mixing within the room. Supply fixtures such as ceiling diffusers, wall grilles, linear diffusers, and others, are designed to cause extensive churning within the room. This allows the fresh, tempered air to mix with the existing air within the room to provide overall comfort. The fact that most office buildings are reasonably comfortable attributes to the success of the mixing action of the supply air systems.

Supply and Exhaust Interface

As previously mentioned, supply ventilation is required to provide replacement of the air that has been exhausted. However, the supply and exhaust work on two different principles. The exhaust system controls laboratory contaminants by gently drawing air away from the source and discharging it outside. The supply system provides control of environmental conditions by thoroughly mixing the air within the room. These two concepts are in direct contradiction. If the supply system is causing significant churning at the face of the exhaust hood, the exhaust hood must overcome the influence of the supply air system in order to capture the contaminant. Consequently, the manner in which the supply air is distributed within the laboratory can be as significant as the method in which the air is exhausted.

Balance

Under normal conditions, it is desirable to have approximately the same supply and exhaust for each laboratory unit. This will provide balance to the area preventing significant airflow through doors, window, wall penetration, and other openings.

In some instances, it is highly desirable to provide intentional imbalance. In these cases, it may be possible to use the ventilation system to provide additional control.

When more air is exhausted from a laboratory than is provided through the supply air system, the room is "negative". When this occurs, air will flow through openings and there will be a general migration from the corridors into the laboratories. In the event that the first line of control, the local exhaust ventilation, does not provide adequate control or a spill or accident occurs, any contamination generated would be isolated to the laboratory. Under a significant release or spill condition, the spill control team can use the corridors to stage entry to the laboratory for clean-up. In less significant releases, the negative condition will maintain the contaminant within the generating laboratory without additional exposure to personnel.

On the other hand, some conditions may require the isolation of the laboratory from other activities. For example, analytical laboratories providing trace analysis, sterile laboratories, some quality assurance laboratories within a production facility and other reasons. Under these conditions, more supply air would be provided to the laboratory than is exhausted. This would cause a "positive" condition with air spilling from the laboratory into the corridors and adjacent spaces.

Under most conditions, the requirement for either positive or negative is not excessive. Supplying slightly more or slightly less supply air will achieve the desired results. Very rarely is it required to maintain a specific static pressure differential between the laboratory and the corridor or between laboratories and other laboratories. If conditions warrant, maintaining a high static pressure differential (here high could be 0.05 to 0.1 in. water gauge) may make it necessary to go with special construction of the laboratory and airlock entries. If a pressure differential of 0.05 in. of water is to be maintained and the door is open, considerable volumes of air would be required. The airflow through the opening would be over 500 linear ft/min. For a standard door, approximately 20 ft^2, 10,000 ft^3/min would be required to maintain the static pressure.

Balance within a laboratory becomes much more difficult when the supply and exhaust systems are manifolded. With multiple exhaust pick-ups and/or multiple supply fixtures, the balancing process becomes an iteration process. Consequently, precise balance can be very difficult to achieve in a large complex building. However, in most instances, precise balance is not required.

Return Air

In many instances, it is desirable to return a quantity of the air supplied to a laboratory back to the air handling system. This is the way that most HVAC systems for commercial and residential application are designed.

A large majority of all ventilation systems return a significant portion of the supply air to the air handling unit. This design is standard for HVAC systems and is widely applied to commercial, residential, and other occupancies.

In laboratories, it is not appropriate to automatically return significant quantities of air. Within many laboratories chemicals are used which can be inadvertently released into the laboratory area and not captured immediately by the local exhaust systems. If the room is provided with return air, this contamination can find its way back to the air handling system and then be distributed throughout the laboratory. The consequences of the inadvertent return of chemicals to the air handling unit will depend considerably upon the nature of the chemical used. In some instances, inappropriate return of air can cause a health or nuisance odor problem.

In my experience, the advisability of returning exhaust air is highly situational dependent. In situations where the exhaust requirements are quite high, the question resolves itself. In situations where the contaminant release is highly toxic and there is a reasonable potential for release of significant quantities of the toxin, it is not advisable to return air from the laboratories. However, when the toxicity of the material is quite low or the probability of incidents is low, returning laboratory air to the central system should be an acceptable design. Since programs in laboratories change, and a laboratory with minimum potential for toxicant release may present a greater hazard as the mission changes, choosing the latter course could later prove to be unwise.

Cross Contamination

Cross contamination describes a situation where a release occurs in one laboratory and exposures occur in another. If the laboratory in which the release occurs is positive relative to the second laboratory, potential contamination can flow from the first laboratory into the second, resulting in exposures. In other cases, the supply system is designed so that a portion of the laboratory air is returned to the air handling units, mixed with the outside air and supplied back to several laboratories.

In many laboratories, significant exhaust ventilation is required by the laboratory hoods. In these cases, all of the supply air to the room is drawn through the exhaust system and discharged outside. If all the rooms on the system have greater exhaust requirements than supply requirements, the system will operate on 100% outside air.

For laboratories which have a return air plenum common to several laboratory units, there is a possibility of transfer of contaminants from one laboratory to another through the return air plenum. For example, if Laboratories A and B share a return air plenum and Laboratory A is positive relative to the plenum and Laboratory B is negative relative to the plenum, it would be possible for air from Laboratory A to migrate through the plenum return system and enter into Laboratory B. When this occurs, simultaneous with a release of contamination within Laboratory A, unacceptable exposures may occur.

Re-Entry

When exhaust air is discharged from a building, it can be drawn back through the air intakes for the supply air system. This is frequently observed in laboratory designs and is accountable for a vast majority of the odor episodes which occur within laboratories. Unfortunately, where odor episodes occur, more significant health exposures can simultaneously occur.

Airflow patterns around a building can be quite complex. As shown in Figure 6.1, the airflow patterns on a relatively simple building can make the location of supply and exhaust systems that preclude the possibility of re-entry impossible.

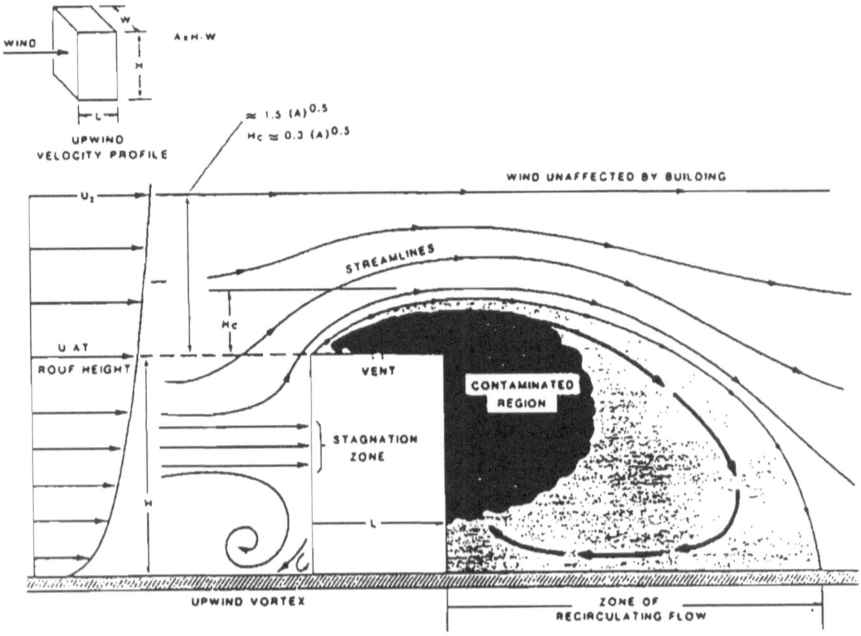

Figure 6.1. Centerline flow patterns around rectangular building.

For multiple storied buildings or buildings located in the vicinity of other buildings, the entire roof may be under the influence of the downwash caused by the adjacent buildings under some meteorological conditons. If there are multiple discharges on the roof, the churning action caused by the downwash will draw air discharged from the laboratory toward the air intakes. This will result in re-entry.

Elimination of all re-entry within a building is probably not feasible. Since significant dilution occurs as the exhaust air migrates toward the air intake, complete elimination of re-entry is not warranted. However, in some instances the toxicity or nuisance value (odor) of the exhaust streams may warrant a reduction in potential re-entry. This can be achieved by providing a collection manifold into which the critical systems can be discharged. Several systems will mix together, resulting in a significant dilution and a single exhaust stack which may be more easily located to minimize the re-entry potential.

4. SUMMARY

In summary, the ventilation systems, both supply and exhaust, play a significant role in the protection provided to chemists and/or technicians in laboratories. Exhaust ventilation, both local and general, provide an avenue for contamination to exit the building, thereby reducing potential operator exposure. The supply system is required to provide adequate replacement of exhaust air and establish creature comfort. The design is critical since poor design of exhaust air can increase the exposure to operators and decrease the efficiency of the ventilation system. Moreover, adequate design of the supply system can provide for a positive or negative condition in the laboratory, providing a secondary level of control.

In some instances, the return of laboratory air to the air handling system, and subsequent distribution throughout the laboratory, is an acceptable practice. However, in other cases, return of air from the laboratories can cause significant health and/or odor episodes and are highly undesirable.

Since the exhaust system does not disappear as the duct passes through the roof of the building, significant concern exists related to the manner in which the air is discharged. Location and/or design of exhaust stacks will result in significant re-entry with increased exposure and nuisance consequences.

CHAPTER 7

Basic Components of Laboratory Plumbing Systems

Henry Koertge

1. INTRODUCTION

Plumbing is a system of pipes, fixtures, and other appurtenances located within a building which transport and distribute water and other fluids including gases from a central source. Plumbing also includes the piping and other accessories which are used to collect and convey the used liquids to a single location outside the building or by which the used fluids are separated for holding and means of treatment/disposal other than through the sanitary sewer system. Laboratory plumbing may include piping systems for cold water, hot water, deionized water, distilled water, natural gas (or LP gas) as well as specialty laboratory gases, vacuum lines and compressed air lines. Plumbing is also the art, trade, or business involved in the design, installation, and maintenance of the plumbing of a building.* Plumbing may also be used as a verb as in "to plumb" a building.

2. SUPPLY

It is important that the user of the laboratory be consulted by the person who is in charge of the layout and design of a laboratory. It is particularly important for the architect/engineer to know what qualities and quantities of water are necessary and where within the laboratory such water is to be delivered.

The water supplied to a laboratory building or to a laboratory portion of a

* Babbitt, H. E., *Plumbing*, 3rd ed. (New York: McGraw-Hill).

building may be from either a public or a private source. Knowledge of the quantity available in terms of both flow and pressure is necessary in order to determine whether or not there is sufficient water for the facility. In some cases minimal supplies may be augmented through the use of on-site pumping and storage facilities. The water will not only be used for drinking and normal laboratory use but should be satisfactory for fire protection, safety showers, eyewash facilities, and the washing and cleaning of laboratory glassware and equipment as necessary.

Except for the most elementary chemical laboratory, the quality of the water supply may be extremely important. The provision of a complete analysis of the water for chemical and physical properties would seem prudent in most cases. Such characteristics are necessary in order to determine the need for and methods of in-house water treatment such as deionization and distillation as well as to aid in the selection of the best materials for the conveyance of treated water. Particularly in smaller laboratories it may be more economical from a long range annual operating cost standpoint to purchase bottled deionized and/or distilled water rather than to provide for such treatment and separate piping systems.

3. TREATMENT

Depending upon the nature and the size as well as the economics of laboratory operation, it may be necessary to provide for in-house water treatment. At times the hardness of the water supplied may be such that zeolite softening is appropriate. When locally available zeolite rental units or those available through hardware stores are not of sufficient size, slightly larger units such as those which may be used in industry or for small public water supplies may be necessary. The next step up in water quality would be to provide for deionized (demineralized) water. In-house treatment could involve either the relatively easily operated anion and cation exchangers (demineralizers) or maybe the somewhat more sophisticated type of equipment which utilizes the reverse osmosis process. In some cases, distilled water systems are installed for laboratories normally on a laboratory by laboratory basis and not for an entire wing, floor, or building. For certain specific laboratory applications it may be necessary to consider additional types of water treatment including activated charcoal filtration for the removal of chlorine and/or organics. It may be wise and helpful to obtain advice from a water chemist or environmental engineer for the water treatment selection and design.

In all cases it will be necessary to select and install piping systems most appropriate to the type of treated water conveyed in those pipes. Water treatment, in many instances, will result in a corrosive water which would require a more resistant pipe material.

4. PROTECTION

One of the biggest potential health problems as well as potential quality control problems, exists when there are conditions which would allow fluids contained in one piping system to enter into and contaminate the contents of another piping system. Thus one basic, necessary function of laboratory design is deciding on one or more methods of cross-connection control. A cross-connection has been defined* as any physical connection or arrangement between two otherwise separate piping systems, one of which contains potable water and the other of unknown or questionable safety, whereby water may flow from one system to the other, the direction of flow depending on the pressure differential between the two piping systems. Another source** has provided a similar description except that it suggests that the physical connection may be either direct or indirect and further explains that an indirect connection consists of a gap or space across which undrinkable water can be sucked, blown, or otherwise be made to enter the drinking water supply while a direct connection consists of a continuous conduit between the two piping systems.

Although it is the primary purpose of cross-connection control to prevent the contamination of a drinking water supply, it may be just as important to prevent the contamination of systems containing deionized water, distilled water or any other fluid contained in a separate laboratory piping system. It should also be noted that the contaminating fluid may not necessarily be contaminated water but could be any liquid chemical (including organics), gaseous in form, and/or be essentially nontoxic. However, these still may present problems when connected to a water piping system for which the contents must be controlled from a quality or potability standpoint. It should be noted that the contaminating/unwanted materials may originate not just in other piping systems but may come directly from compressed gas cylinders as well as unpressurized liquid containers such as sinks, drains, buckets, pails, beakers, etc.

"Backflow" occurs when cross-connections exist and when an abnormal pressure differential is created causing a reversal in the direction of flow in the piping system to be protected. When the lower pressure creating a backflow condition is less than atmospheric (vacuum condition), the backflow condition is known as back syphonage. Specific detailed methods required to prevent backflow are contained in plumbing codes.

The only known positive way to prevent backflow into a piping system from either another piping system or fluid container, is to maintain an adequate air gap between the two. Where it is essentially impossible or virtually impractical to maintain an air gap between the two systems, some sort of mechanical device needs to be relied upon. Various mechanical backflow prevention units with varying degrees of reliability include but may not necessarily be limited to (1) atmospheric vacuum breakers, (2) double check valves, (3) double check valves with atmospheric vents, and (4) reduced pressure backflow preventers. These

* *llinois Plumbing Code*, Illinois Department of Public Health, Springfield, Illinois, 1986.
** Babbitt, H. E., *Plumbing*, 3rd ed., (New York: McGraw-Hill, 1960).

backflow preventers are allowable by plumbing codes only in certain specified situations and it should be emphasized that the devices must be installed correctly in order to function to the limits of their capabilities. As an example, an atmospheric vacuum breaker must be located above the highest point of use of the downstream flow, not exposed to any downstream pressure higher than atmospheric, and must be installed on the outlet side of the last control valve. It is also conceivable that a vacuum breaker can be improperly installed so that the water flow through the device is in the wrong direction though most such mistakes would immediately be evident when the system is tested.

It should be especially noted that the reduced pressure backflow preventer which is accepted for use by several codes for various applications is also subject to failure as is any mechanical device. They must be tested on a semiannual or annual basis which requires two in parallel so that the extra device can be in operation during the testing period.

By far the safest and perhaps best design for the protection of the quality of the drinking water supplied to the building or to a laboratory portion of a building is to install a completely separate piping system for laboratory water usage. While this may be required in certain localities, it may also be the method dictated by the economics of the project particularly when compared to the option of providing a vacuum breaker at each outlet. A separate system is created by splitting the water supply with one piping system serving drinking water fountains and restrooms and the other for laboratory usage. The latter is directed through an air gap into a surge tank where it is pumped to a pressurized storage tank from whence water is piped directly to the laboratories. In this manner there is absolutely no way that contaminants can enter the potable water supply so that control of cross-connections to the laboratory system with vacuum breakers is unnecessary. It should be noted, however, that this unprotected lab system allows backflow into and contamination of the water if cross-connections exist and backflow conditions arise. Therefore, the separated laboratory piping system should be considered as being nonpotable and so labeled at all outlets (faucets).

5. PIPING

Pipe installed for the conveyance of liquids and gases for laboratory purposes must be selected depending upon the type and quality of fluid being conveyed and may also be regulated, specified, or limited by plumbing codes. Characteristics of pipe material would include but not necessarily be limited to: (1) durability, (2) ease of installation, and (3) anticorrosivity. Materials most often used are galvanized steel, copper, plastic, and aluminum. In order to minimize future operation and maintenance problems, it is necessary to insist upon the best of workmanship in the installation of the piping systems. The installation of some piping materials requires extra care and skill in order to minimize future failures of the system. The finished system must be hydraulically tested. Sloppy techniques during the installation of some piping materials may also result in a system which may be contaminated with materials such as cutting oils which are virtually impossible to remove after the installation is complete. Final acceptance of the system, therefore, may hinge on tests for such contaminants as well as the testing of bacteria for sanitary quality purposes.

In order to facilitate operation and maintenance procedures, it is recommended that the piping systems be color coded in accordance with *Scheme for the Identification of Piping Systems,* ANSI A-13.1-1981 (R1985) or, alternatively, comply with local building/plumbing code requirements pertaining to distinctive pipe labels or colors. Knowing the contents of each pipe is extremely important for future additions to the piping systems. Equipment such as ice machines and vending machines have been inadvertently connected to air conditioning chilled water lines and hot water heating lines resulting in problems ranging from palatability to severe illness.

6. FIXTURES

The fixtures to be installed in the laboratory include but are not necessarily limited to sinks, cup sinks, faucets, safety showers, and eye wash facilities. Sinks and cup sinks must be constructed of materials which are resistant to the types of materials that would come in contact with the interiors of the sinks. Guidance for the installation and performance of both safety showers and eye wash facilities is contained in *Emergency Eye Wash and Shower Equipment,* ANSI Z358.1-1981, which suggest that each type of unit be in accessible locations that require no more than 10 sec to reach and that the travel distance be no more than 100 ft from the hazard.

7. WASTE

Water is used in the laboratory for various functions such as washing, rinsing, cooling, and is directly incorporated into various mixtures and solutions with chemicals utilized in the research/experimentation. These processes create a wide variety of contaminated liquids. Compliance with present hazardous waste disposal regulations would prevent the discharge of some of these wastes directly to the sanitary sewer system. Certain highly reactive, corrosive, and/or toxic mixtures resulting from laboratory experimentation must essentially be removed from the normal waste disposal process and be separately stored, treated, neutralized, and/or transported off site for separate disposal. Some mixtures resulting from in-house batch treatment may be discharged to the drain system so long as they exhibit no characteristics which may be detrimental to or react with the drain/vent materials, react with other discharged materials, or be in conflict with sanitary sewage quality criteria as may be limited by local ordinances. Such criteria are established to help assure the treatability of the waste received at the local waste treatment facility and are usually enforceable by local ordinance.

8. DRAINS

Drains are installed to receive used water from sinks and convey liquids discharged to floor drains. The drains must be designed for size and grade and must be constructed of materials that are durable and which will resist the chemical and physical abuses of the system. Cast iron, copper, plastic, and glass have been used for laboratory drains. The drain system must include adequate numbers of clean-outs in appropriate locations as well as to provide for traps. Traps are necessary to prevent sewage odors and volatilized chemicals emanating from the waste water from entering the laboratory and causing health/nuisance problems. The drying out or loss of seal in traps can be minimized by specifying traps with deeper seals, small surface areas, and greater volumes. On some occasions it might be appropriate to install some sort of mechanism which would automatically add water to each drain/trap in order to eliminate the periodic servicing of them manually. Caution should be exercised, however, so that the installation does not create a cross-connection. In addition to establishing a routine of adding water to all seldom-used drains, it may be prudent to add small amounts of nonhazardous, degradable oil to reduce the evaporation from the trap seal. The drainage system, like the supply system should be tested prior to acceptance and normal usage.

Special consideration should be given to the installation of floor drains. The installation of floor drains for any reason, including those normally located under emergency showers, might not be advisable even though they may be required by code. One reason for caution is the likelihood that the materials from spills, breakage, or from contaminated persons would be hazardous and, therefore, should not be discharged to the regular sanitary sewer. It may be more environmentally sound to allow accumulation of liquids on the floor so that these chemicals can be cleaned up, neutralized, or separately disposed. An additional negative factor involving installation of floor drains is that being seldom used, the traps frequently dry out allowing odors and volatile chemicals to enter the work environment from the building sewer system.

The designer should be wary of code requirements for "generic neutralizing tanks" or basins which in the past have been required for drains from "all laboratory buildings." Because of the variety of types and strengths of wastes that emanate from any laboratory, the installation of an underground tank containing nothing more than limestone is of extremely minimal and therefore doubtful value. Now with the hazardous waste disposal regulations which for all practical purposes require the separation of all significantly detrimental materials, the need for the installation of a "neutralization tank" has virtually, if not completely, vanished. These hazardous chemical regulations have also minimized the need for chemically resistant drain materials such as glass.

9. VENTS

In the waste portion of the plumbing system, the vents play important and yet sometimes misunderstood roles in the overall system. Vents are necessary to equalize the air pressure in the drain system caused by the flow of liquids or for other reasons. This pressure equalization prevents the loss of the seal in the traps (which as previously noted prevents odors and volatile chemicals from entering the building from the sewer system). In order to create the equalization of air pressure, vent pipes are installed leading to the roof so that noxious, volatile, or other gaseous materials emanating from the waste water are discharged to the atmosphere. Where main or house traps are not required, this venting system also aids in the ventilation of the sewers outside the building. One of the main health and safety factors from a design standpoint is to properly locate the outlet end of these vents relative to fresh air inlets to the building. While it is unlikely that unhealthy or nuisance conditions exist other than in close proximity to the outlet end of the vent, some consideration may also be given to the location of the vent outlet on the roof as it may pertain to persons working on the roof as well as the location relative to the air inlets for adjacent buildings.

Vents must be constructed of materials which are durable and resistant to the chemical effects of the waste and where practical should be accessible for visual inspection. The completed systems should be tested prior to normal usage.

CHAPTER 8

Do's and Don'ts in Providing for Storage of Chemical Wastes in the Laboratory

Peter Ashbrook

Storage and handling of hazardous wastes generated in laboratories is a relatively new area of concern within the broader context of laboratory design. For years, it was assumed that all kinds of nasty chemicals would be disposed down the drain, and that the laboratory drainage system should be designed accordingly. Thus, in reviewing old building plans, one may find drainage lines from laboratories identified as "Toxic Waste", "Hazardous Waste", or a similar designation. A common engineering practice was, and still is, to run this effluent through a limestone trap, which would presumably neutralize any acids that might be present. However, any provisions for hazardous waste management beyond the use of limestone traps has been highly unusual in the design of new or remodeled laboratories.

With the passage of the Resource Conservation and Recovery Act in 1976 and closer scrutiny of the discharges to wastewater treatment plants, pressure has been building to eliminate the use of the drain for disposal of waste chemicals. Persons involved in laboratory design planning would be well advised to spend time determining what can be done in terms of design to facilitate sound management of waste chemicals. This chapter will present an overview of potential hazards, planning considerations, provisions for storage of waste chemicals, the role of waste minimization, and comment on common practices — some good and some not so good.

1. POTENTIAL HAZARDS

In general, the hazards of waste chemicals in the laboratory are similar to the hazards of the same chemicals when used in the laboratory before they become wastes. These hazards, according to the Environmental Protection Agency (EPA), include ignitability, corrosivity, reactivity, and toxicity. For purchased chemicals, one is able to obtain a Material Safety Data Sheet (MSDS), which may, depending on the individual who prepared it, contain useful information about its hazardous characteristics. Chemical wastes, on the other hand, frequently are mixtures of chemicals, and rarely are MSDS's available. Therefore one must take special care in the handling of chemical wastes.

The most common potential hazards related to storage of waste chemicals are

1. Mixing incompatible wastes in the same container
2. Spills, most commonly through breaks or leaks in the container
3. Fire and safety hazards due to improper storage, poor housekeeping, and/or poor ventilation

Mixing of incompatible wastes would appear to be an easy hazard to avoid. However, in a laboratory situation there is such a wide variety of chemicals that it is difficult for an individual to be aware of all the potential incompatibilities. Compounding the complexity due to the variety of chemicals used, is the fact that laboratories are often used by many different persons. In the academic setting, particularly, there is a tremendous turnover in the number of persons using a laboratory. Common incompatibility reactions include:

- Heat generation from acid/base reactions
- Explosions and heat generation from polymerization reactions
- Fire from the mixing of water reactives with wastes containing water
- Generation of toxic fumes

Spills of chemical wastes typically occur from breakage of glass containers or corrosion through metal containers. Empty solvent and acid bottles are often used by laboratory workers for storage of wastes. It has been my experience that the use of poly containers is usually preferable because poly containers are less likely to break. Metal safety cans are often used to store waste solvents; however, if the solvents include any halogenated compounds, there is a great potential for corrosion. Halogenated compounds when mixed with water may dissociate to form an acid (such as HCl). When this happens, one finds the leak occurring in horizontal lines at the water layer.

Housekeeping practices can greatly affect safety regardless of how good or bad a laboratory is designed. Common housekeeping problems include:

- The storage of too much waste in the laboratory
- Storage of waste containers where they are likely to be disturbed

- Storage of incompatible chemicals in close proximity to each other
- Use of the same fume hood for both lab work and waste storage
- Unlabeled or inadequately labeled containers of waste
- Storage of waste chemicals in unsealed containers
- Storage of waste chemicals for too long (e.g., peroxide formers)

The above hazards are significant enough that attention should be spent on preventing them. In addition, it should be anticipated that accidents will happen. Planning should therefore include response procedures in the event of an accident. Frequently, injuries occur in response to incidents even when the personnel present when an incident occurs escape injury.

2. PLANNING FOR HAZARDOUS WASTE STORAGE IN THE LAB

Avoidance of potential hazards related to the storage of hazardous wastes can be facilitated by good laboratory design. The following questions should be answered in the planning stages for the design of laboratories:

1. What kinds of chemical wastes will be produced?
2. How much of each type of waste will be produced and at what frequency?
3. What types of containers will be used to store waste chemicals in the laboratory?
4. Where will waste chemicals be stored in the laboratory?
5. How much storage space will be required for each hazard class of chemical waste?
6. What are the responsibilities of the laboratory workers with respect to the handling of waste chemicals?
7. How often will waste chemicals be removed from the laboratory?
8. What will laboratory personnel do if waste chemicals cannot be removed from the laboratory in a timely fashion?
9. Is storage or working space required for redistillation of solvents, redistribution of partially used containers of chemicals, treatment of wastes, or other waste minimization procedures?
10. Under what conditions, if any, will waste chemicals be allowed to be disposed down the drain?

The answers to these questions will have a significant impact on design considerations. Fortunately, many of these concerns will be addressed when consideration is given to the intended laboratory uses. However, the potential hazards of waste chemical management should be addressed as a separate issue to ensure that these potential hazards receive adequate attention.

3. COMMON PROBLEMS/ISSUES AND SOLUTIONS

As with most laboratory safety issues, prevention of hazards can be accomplished through a combination of good working practices and good laboratory design. Frequently, the two go hand in hand. The problems discussed below will focus on design aspects, but will inevitably contain mention of good working practices. This section will present some common issues or problems faced in laboratory design and common solutions. The reader should be aware that often there are several satisfactory solutions. It is hoped that the examples and information presented in this chapter will prepare readers for analyzing the issues to be addressed in designing their particular laboratories.

The Chemical Stockroom

Larger institutions will usually have one or more central chemical stockrooms. What better place to store waste chemicals than a place where chemicals are already being stored? Chemical stockrooms may be satisfactory for storage of waste chemicals. However, I have observed many chemical stockrooms that have too little space, unsatisfactory storage cabinets, inadequate ventilation, and/or little or no supervision. If your stockroom is one of these, think twice before adding to the problems by putting it into use for waste chemical storage.

Additional space in the stockroom can be provided by reducing the size of the inventory of new chemicals in storage and the size of containers in which chemicals are purchased. Frequently, chemicals are purchased in larger quantities to get price breaks. Later it is discovered that too much was purchased and the excess must be disposed. Make sure that both usable chemicals and waste chemicals are inventoried frequently. Although ready access to the stockroom by laboratory workers may facilitate work being performed in a timely fashion, safety considerations dictate that stockrooms should have restricted access and be managed by well-trained attendants.

Many laboratories have their own individual stockrooms in a small room connected to the main laboratory. The above considerations apply to these small stockrooms. It is imperative that responsibility be given to a single person for maintenance of the stockroom.

The Centralized Chemical Waste Storage Room

One solution that is frequently considered for storage of chemical wastes is the designation of a centralized waste storage room in a building. In my opinion, this solution is unacceptable unless the room is limited to access by a single individual. If everyone has access, no one is in charge and wastes will accumulate out of control. In the vast majority of cases it will be far better to provide storage space for hazardous wastes in each lab and put the room designated for storage of waste chemicals to use as another productive laboratory.

Use of the Sewer

It was mentioned earlier that the sewer used to be used for disposal of waste chemicals. In many laboratories, this practice continues. I do not advocate the elimination of the use of the sewer; however, if the sewer is not used responsibly we may find that this option will be taken away from us.

The first step in the responsible use of the sewer is to design the laboratory to encourage the optimum methods of handling waste chemicals. As with many of the issues in this book, good operating procedures play an important role as well. If, on the other hand, the attitude during the design stage is one of letting the laboratory workers solve their own waste disposal problems when they arise; it is likely that many waste chemicals will end up going down the drain for lack of an alternative.

Going one step further, I have seen proposed building design plans in which the use of the drain was encouraged. One proposal was to provide separate drains for waste solvents and waste acids, and to provide holding tanks so that pretreatment could be provided before release to the sanitary system. To my way of thinking, this idea was a classic case of looking good in theory, but in actuality would produce total disaster. The first problem is that even after segregating solvents from acids, there would still be a great potential for incompatibility problems. Second, who is going to train the laboratory workers to make sure that the proper drains are used at all times? Last, does one really want to encourage the mindset that the drain is the best disposal option for laboratory waste chemicals? Note that if this holding tank idea is pursued, the tanks may be subject to permitting under the U.S. EPA hazardous waste regulations.

One may wish to design a laboratory so that the drains are acid resistant because it is very difficult to control what will go down the drain. However, this situation is much different from designing a laboratory to encourage wastes to go down the drain.

Waste Minimization

The current hot area of hazardous waste management is to push waste minimization—and rightly so. In fact, there have been a number of serious proposals to mandate waste minimization for all generators of hazardous wastes. Whether such mandates will eventually affect laboratories is unclear, but the possibility certainly exists.

Two major ways to minimize wastes are to prevent its generation in the first place or to recycle the wastes that are produced. Here again, good laboratory design can encourage waste minimization. Perhaps the most common example of waste minimization is the recovery of solvents by redistillation. Redistillation can save money both by avoiding disposal costs and by avoiding purchase costs of new material. Often, the products of redistillation are of better quality than the original purchased material. If redistillation is to be considered as an option, there must be adequate space. If space is inadequate, one can bet that this option will not be considered by the laboratory users.

Many types of treatment to reduce or eliminate the hazards of waste chemicals are available. These treatment procedures can reduce the need for sending waste chemicals for treatment and/or disposal off-site. Again, if these procedures are to be pursued, sufficient space with the right amenities must be provided.

Storage of Wastes

Waste chemicals are frequently stored under sinks or in open space on the floor. To prevent the disturbance of these containers, they should be stored in suitable cabinets. Many laboratories do not have suitable storage space for their new chemicals, let alone space for waste containers. The best opportunity to attack this problem is when new or remodeled laboratories are being designed. Two factors to consider are to (1) make sure all storage cabinets are ventilated and (2) make sure that there are enough storage cabinets so that wastes can be segregated from usable chemicals and wastes of different hazard classes can be segregated. In the case of storage cabinets for flammable solvents, one must take care that the provision of ventilation does not create the potential for an explosion hazard.

4. SUMMARY

Generally, the considerations for storage and handling of waste chemicals are similar to those for the storage and use of new laboratory chemicals. A well thought out laboratory design will facilitate optimum handling methods for waste chemicals and encourage the safe storage of wastes within the laboratory.

Section III: Ventilation and Fume Hoods

CHAPTER 9

Working Together to Design Safe Laboratories: Ventilation, A Consultant's Perspective

Earl L. Walls

I had my first exposure to the programming and design of a laboratory in 1951. Thirty-seven years, and then as it is today, ventilation is the most perplexing of problems we face in the design and operation of laboratory buildings. On the average, ventilation systems are contributing about 30% of the cost of today's laboratory buildings and about 80% of their problems. Though to be perfectly candid about it, much of the latter is due to the lack of knowledge on the part of the lab users about the systems and how to use them.

Engineers and architects designing today's laboratories are faced with three somewhat divergent demands. Meet all the safety requirements, e.g., fume hood face velocities and air exchange rates; conserve energy; but build it cheap and minimize upkeep costs. In some instances there is another demand to make it so fool proof that the lab occupant does not have to assume any responsibility. Simple?! Hardly!

Though we are forced to follow the established guidelines in terms of the provision of fume hoods and their consequent face velocities, I, for one, take issue with many of these. Or, at least, how they are interpreted. I believe we have to revisit the basic parameters. Really redefine what is safe and what is not safe, without using the copout of putting everything in hoods to make it safe. That is certainly the easiest thing to do from the viewpoint of the industrial hygienist, and for that matter it is the easiest thing for engineers to do. It is easy, but it's sure not cheap or energy efficient.

Consider, for instance, that in round numbers in the U.S. every 200 cfm exhausted through a fume hood equates to about 12,000 btu, or 1 ton of refrigeration. An 8-ft wide hood, wide open, with a face velocity of 100 ft/min consumes between 1750 and 2000 cfm — 8 to 10 tons of air conditioning per hood at an installed cost of 5000 to $6000/ton. But the real questions remains — is it safe? Maybe. Maybe not. Certainly not if the hood has supply air impinging on its face at 80 ft/min. Certainly not if the hood users do not carry out their work at least 6 in. back of the sash line. But, with all things right in that regard, is this type of protection really necessary? Or should a more protective method be used, e.g., the glove box. It boils down to defining the degree of hazard that my friend Gerhard Knutson likes to discuss so much. To be empirical about it, however, takes time and money that to date we are not willing to invest.

That being the case let me address quickly what can and is being done toward trying to meet the goals of safety, energy conservation, and cost.

Using whatever face velocity that is considered safe, one obvious method is to reduce the face opening of the hood. This can be done by restricting the height that a vertical sash hood may be opened; use a horizontal sash hood that restricts the opening while providing access to the top of the hood; or the method I prefer which is the combination of horizontal sash in a vertical rising sash. These methods result in reducing the air consumption in an eight foot hood to about two fifths of the full open quantity. The horizontal sashes also provide a built-in safety shield. That's about it for the "state of the art" hoods.

Hoods are only a component of the overall ventilation systems, however, so here are a few ways we can incorporate them today in striving to meet the three goals.

1. Couple the hood to an exhaust system control, either dedicated or combined, that is interlocked with the supply system control as a variable volume/constant volume/reheat system (Figure 9.1). In this system, with the hood in the "off" mode, the room thermostat calls for a variable volume of air to offset heat gain and the hood controller tracks on the supply to maintain the air balance in the room. In this manner the hood is never really "off". When the hood is in the "on" mode, the hood controller goes to full open to assure proper face velocity and the supply controller goes to full open to balance the space while a reheat coil maintains temperature control. This system is relatively easy to maintain.
2. The latest approach is somewhat similar but a lot more complicated and more difficult to maintain (Figure 9.2). In the simplest form, this approach uses a controller on the hood that maintains a constant face velocity regardless of the sash position. This is done either through a control damper on the exhaust system or with a variable speed exhaust fan. This requires then that the air supply be varied to balance the air being exhausted, but in order to maintain temperature control and provide for general exhaust another system has to be in place. So we have a hood balance system and a room balance and control system. The latter may be a constant volume reheat, or a variable volume reheat system. Time does not permit a full dissertation on this system but any building being considered today should look at it.

As to further energy conservation, I am a proponent of the use of combined

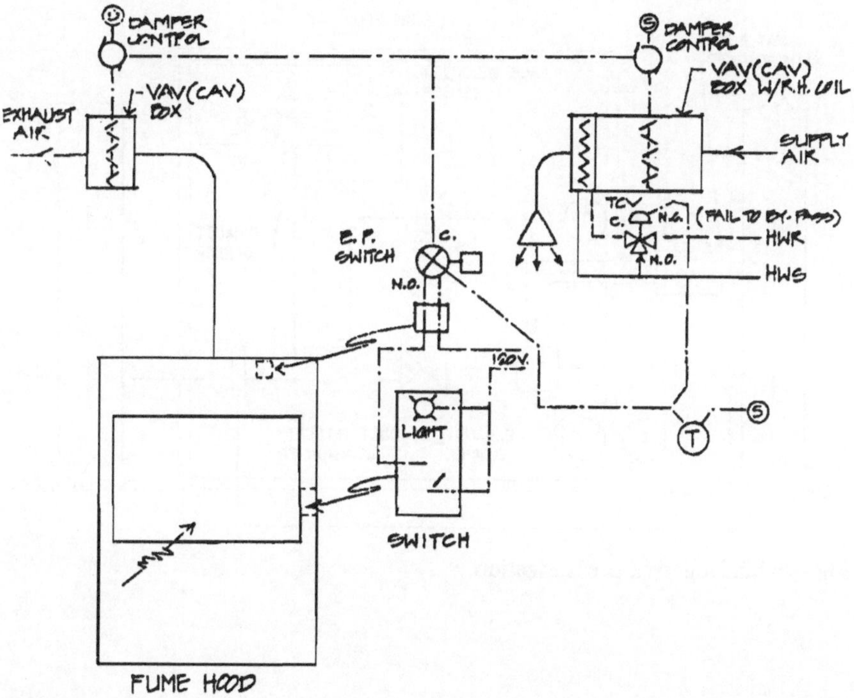

Figure 9.1.

exhaust systems, i.e., one system exhausting numerous hoods, to facilitate the use of thermal recovery components. The use of such components on individual exhaust units would never be cost effective.

That gets me to the last consideration that time permits. The user! We have done a very poor job of teaching the laboratory workers how to use what has been given them in terms of laboratory ventilation. No system, regardless of how well conceived, designed, and built will be safe if improperly used. Any emissions from procedures carried out within 6 in. of the back of the sash are almost bound to find their way into the breathing zone of the user. Pouring from one vessel to the other while holding them at or near the hood face will likely get it — in the face.

Just one brief note about my earlier remarks regarding exposure. Obviously, we must be safe, but just what is safe in the laboratory? Consider for instance that a worker in a chemical plant manufacturing a monomer is limited to so many parts per million per 8-h day. In the laboratory we seem to have the attitude that one whiff is deadly. For goodness sake, if it is, it sure as the dickens does not belong in the fume hood — it belongs in a glove box. All I suggest is that we spend more effort in truly defining the problem, not in throwing cfms at it.

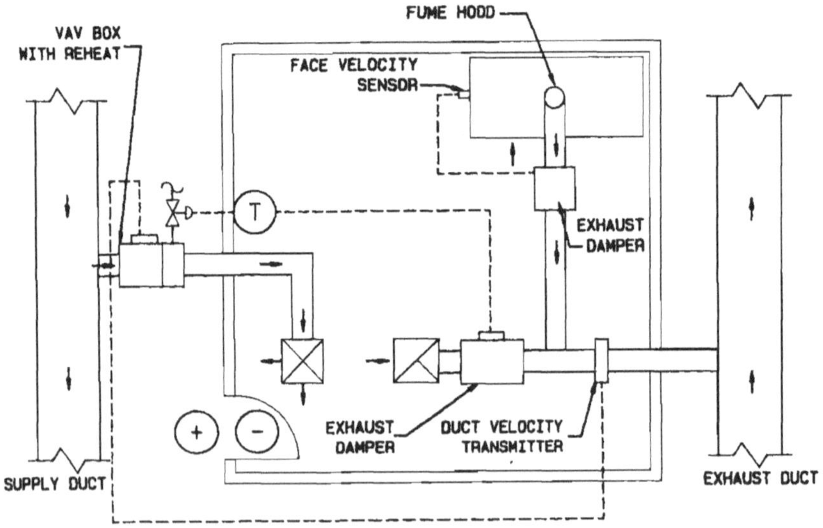

Figure 9.2. Negative pressurization.

CHAPTER 10

Basic Principles of Fume Hood Design and Operation

E. Robinson Hoyle and R. Scott Stricoff

1. INTRODUCTION

This chapter will explore two major elements in regard to fume hood design and operation. The first element is to review the complete fume hood system from the hood to the exhaust stack from a design prospective. The second element will review the administrative controls (e.g., moderating and maintenance) which are necessary to provide proper operation of the hood. This chapter will deal with the balance between the two elements, design vs. administrative controls. It is important to note that a good fume hood design can always be defeated through the lack of adequate administrative controls. Although more difficult, the opposite is true that a poor design can be overcome by a good administrative control program. Consequently, no matter how well intentioned the design, it can always be overcome through human interaction. In order to make these two points, this chapter will be divided into three sections including: design and hood selection, installation selection and testing, and operation and maintenance.

2. DESIGN AND HOOD SELECTION

There are many different types of laboratory exhaust ventilation systems. These include chemical fume hoods with different designs (e.g., introduction of supply air and location of baffles), biological safety cabinets with recirculation of the laminar flow or total exhaust of the laminar flow, glove boxes, and many miscellaneous systems including slot hoods, enclosures, elephant trunks, and large walk-in hoods. This chapter will be devoted primarily to chemical fume hoods.

Examples of chemical fume hoods and uses include the weighing and transfer of toxic chemicals, performing analytical tests and procedures, and storage (sometimes improperly) of various items. Further examples of hoods include re-circulation fume hoods with charcoal filters (sometimes referred to as ductless fume hoods), and ventilated boxes.

Biological safety cabinets are designed to provide both worker and product protection with exhaust and laminar flow airstreams. There are three different types of cabinets providing different levels of personnel and product protection ranging from 70% internal recirculation to total exhaust of all cabinet air. Biological safety cabinets come in all shapes and sizes. Different modifications of biological safety cabinets include one built with an incinerator (heated electrical coil) in the exhaust duct to sterilize the exhaust air. Clearly, this is a unique design which is slightly outdated and may present added safety hazards to the employee.

The design issues associated with the hoods described above include introduction of make-up air; design of the air foil; the shape, size, and introduction of ductwork (elbows and transitions) to the cabinet; the location and placement of air cleaning equipment (if any); exhaust stacks (height and placement); and fan location. The first step in designing any system is to understand the ventilation system components. Figure 10.1 illustrates the hood, the ductwork, fan, and exhaust stack. The following discussion of hood system design criteria includes actual hood characteristics and placement in the work area.

The ideal location for a laboratory hood is one in which turbulence at the face of the hood will not be created by factors such as the general supply and exhaust diffusers, employee traffic, open windows or doors, or interaction with other ventilation systems. As an example of the latter issue, we ran across two hoods connected together in a series located across the aisle. In this situation there will be an interaction between each of the hoods irrespective of traffic and face velocity because the hoods are located so close to each other. In this case, one hood was creating a back draft into the other hood due to poor design for the duct connection. In another case, we observed a glove box which was introduced incorrectly into an older chemical fume hood. The transition should not be perpendicular to the duct, but at a 45° angle. Any time a connection is made to a hood or duct, the impact on the face velocity to the existing chemical fume hood should be reviewed and calculated by a qualified individual. Too many elbows and directional changes in the ductwork can create a significant amount of duct turbulence and air friction, thereby reducing the efficiency of the exhaust system.

Additional design criteria for laboratory hoods include: the air cleaning equipment, monitoring systems for filter loading, fan sizing, and duct pressures with respect to working areas. This latter issue involves the placement of the fan and

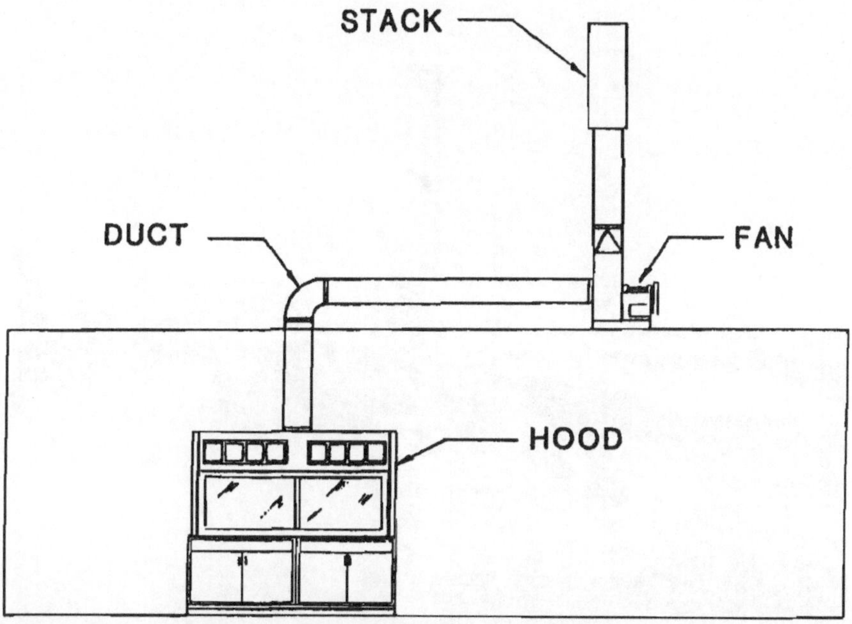

Figure 10.1. Diagram of a basic laboratory hood system.

how it pressurizes the ductwork. Leaky ductwork under positive pressure with respect to the laboratory space will let extremely toxic material in the exhaust airstream circulate back into the lab. When used, the type and placement of filtration equipment is important. When filters are used, we recommend that they be installed in duplicate so the exhaust airstream can be shifted to provide protection during the filter change process. Monitoring systems should be installed before the first filter, in between the two filters, and upstream of the final filter to monitor filter contaminant loading. The location of the exhaust fan is also important in such a system to assure that the filters and adjoining ductwork remain under negative pressure with respect to the environment, to prevent toxic contaminant leakage.

Finally, the exhaust stacks must be located high above the building to assure that the exhaust air is ejected up and away from the building and not reintroduced into the building's supply air. The potential for exhaust air re-entrainment is shown in Figure 10.2. The air outlet for two exhaust systems (left side) are approximately 3 ft from the air supply inlet for the building assuring that there will be a re-entrainment of the exhaust air. The relationship of the discharge stack height to the building height is critical to avoid the re-introduction of contaminated air because of the wave cavity created on the lee side of the building. The wave cavity can bring contaminated exhaust back onto the roof, side or ground level of the building If there is not enough stack height to eject air above the wave

Figure 10.2. Fume hood stack discharges with weather caps. Note air intake in the middle of the picture.

cavity, contaminants can be brought into an air inlet on the left side of the building. The exhaust stack height should be 1.3 or 2.0 times higher than the building to prevent entrainment into the wave cavity.

Exhaust stack height is important, but also capping of exhaust stacks will provide opportunities for re-introduction of exhaust air. Weather caps (Figures 10.2 and 10.3) are unacceptable due to the potential to impact the discharge velocity and force contaminated air back down onto the roof. Exhaust stack designs, which prevent rain from moving back down stacks, include providing a double wall, offset elbow, or offset stack (Figure 10.3).

In summary, proper design issues include understanding the interaction between the exhaust stack and fan location, looking at ductwork and understanding transitions and elbows, looking at hood locations in the laboratory with respect to traffic, window, and doors, and the location and introduction of exhaust and supply diffusers.

3. INSTALLATION SELECTION AND TESTING

Within the past 5 years, several new tests have been introduced to help the fume hood designer understand the important design characteristics of hoods. Examples

Figure 10.3. Examples of weather caps (left) and offset stack design (right) to prevent rain from entering the ductwork.

of these two tests include the ANSI/ASHRAE 110-1985 and the Chamberlin test utilized by the Environmental Protection Agency (EPA). Both of these tests are designed to provide quantitative hood capture information based on the introduction of a contaminant inside the hood and its potential for escape. Both tests introduce gases into the hood at known rates and with the use of a mannequin and detector, measure the escaping gas on the outside of the hood. For example, ASHRAE 110 utilizes a trace gas dichlorofluoromethane or sulfur hexafluoride. This tracer gas is introduced into the hood at a flow rate of 1 to 8 L/min and a detection instrument with a detection limit of 0.01 to 100 parts per million (ppm) (an accuracy of ± 10% at greater than 0.1 ppm, and ± 25% at greater than 0.01 ppm) is used to detect any released gas.

The ACGIH *Industrial Ventilation, A Manual of Recommended Practice* states that "Any well-designed airfoil hood, properly balanced can achieve < 0.10 ppm control level when the supply air distribution is good." Consequently, flow rates can be correlated with the amount that escapes outside the hood. To test the hood effectiveness, a mannequin is placed in front of the hood with a tracer gas ejector inside the hood. The detector measures the contaminants in the breathing zone of the mannequin. The sash can be placed at different heights and air flow is recorded at different times during the testing.

Currently, both of these quantitative tests have some controversy due to their inability to provide consistent data on currently installed existing systems. Early applications of these methods to existing installations have shown that due to variable operating characteristics and the placement of the hoods, interpretation of the quantitative data is difficult. However, the tests appear to have application to evaluating new hood designs. Additional application of the tests and research will help to clarify the quantitative hood tests' role in ventilation design and maintenance programs.

4. OPERATION AND MAINTENANCE

As stated earlier, a well-designed hood can provide inadequate protection if administrative controls are not in place. The design can always be overcome by improper operation and installation. A common problem is placement of air conditioners near fume hoods. In some cases, exhaust contaminants from a hood may be pulled in by a poorly located air conditioner. In other cases, the air discharge from an air conditioner may cause turbulence at the face of the hood resulting in the escape of contaminated air into the room. Hood performance criteria controlled administratively through maintenance programs include examination of internal and external air patterns, the capture velocity, any leakage of plenum sashes or gloves in the case of glove boxes, working area downflow in the case of biological safety cabinet laminar flow, pressure gauges, and routine preventive maintenance on ductwork fans and air cleaners.

Administrative programs for hood monitoring and evaluation include daily visual inspections, quarterly inspections, user training, and routine maintenance. Improper storage of chemicals and equipment is a common and major misuse of hood space.

Daily visual inspections include the review of exhaust slots to make sure they are clear. In many cases, in biological safety cabinets exhaust slots can be adjusted. During the course of adjustment, screws can come loose and vibrate down. In one case, we came upon a biological safety cabinet, for which the laboratory had checked the face velocity and found it to be 100 ft/min. Introduction of a smoke candle into the hood indicated that the 100 ft/min face velocity was in the wrong direction! The rear exhaust slot had vibrated shut providing just the laminar flow to the hood user. Needless to say, the user in this case was very disturbed that the hood was in fact providing no protection.

In addition, the daily visual inspections should include an air flow check for both hoods and glove boxes. This can be done with the use of tissue paper to make sure the air is moving in the proper direction. Daily visual inspections should include review of any pressure gauges to assure that the duct static pressures are proper. The hood user should understand the range of the pressure gauges. In many cases, mangehelic gauges can indicate poor fan operation or overloading of filters.

Quarterly inspections of fume hoods include smoke tests and quantitative measure of face velocity. Smoke tests (Figure 10.4) can include use of a smoke

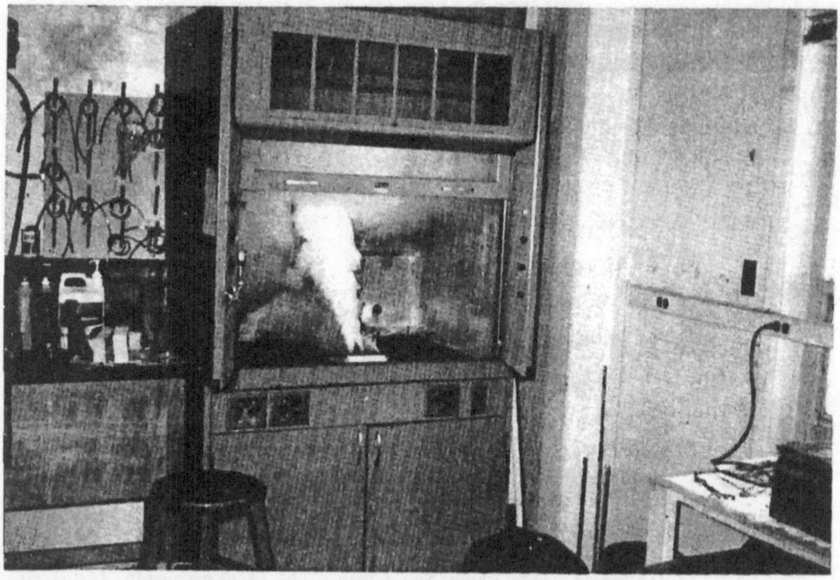

Figure 10.4. Smoke test of fume hood demonstrating good capture of contaminants.

candle or smoke tube. Fume hoods can be reviewed for dead spots and turbulence. Biological safety cabinets can be smoked to examine the relationship between the laminar flow and exhaust intake and any leaks in or around sashes. Glove boxes can be reviewed for glove leakage and leakage of gaskets around the gloves. A smoke tube can be placed in front of the hood and gradually release smoke across the hood face to look at the capture efficiency of the hood. Smoke candles can be utilized to look at air patterns within the hood.

Quarterly inspections should also include the measurement of the face or capture velocity in a fume hood. In a fume hood, the measurement of velocity should be an average velocity using 5 to 6 points across the hood face. Maximum worst conditions should be taken into account during this measurement (sash wide open, blockage of exhaust slots). In addition, the introduction of supply air should be factored into how the face velocity is measured. For the quarterly inspections, some combination of smoke tubes and face velocities tests have to be utilized. Both tests will help to understand a variety of important issues. For example, for many years face velocities of 150 ft/min were thought to be acceptable to protect employees. Recently most guidelines, including the ACGIH, recommend face velocities of closer to 100 ± 20 to protect personnel due to the creation of eddies by high velocities with people in front of the hood. High face velocities can also upset small containers and paper items.

Additional routine maintenance includes review of the exhaust fan for lubrication, belt slippage and deterioration, and fan speed (rpm). Ductwork must be reviewed for corrosion damage, liquid or solid condensate, damper lubrication, and removal of unused hood or duct.

In conclusion, this chapter has briefly gone over design criteria of hoods as well as administrative control programs. The reader should note that these issues brought up are not discussed in detail and a list of references at the end of the chapter provide more detailed information.

References

1. Caplan, K. J. and G. W. Knutson, "Influence of room air supply on laboratory hoods." *American Industrial Hygiene Association Journal,* 43(10): 738-746, (1982).
2. HVAC Systems and Applications Volume, Chapter 30, Atlanta, GA: American Society of Heating, Refrigerating, and Air Conditioning Engineers, (1987).
3. *Industrial Ventilation: A Manual of Recommended Practice* (Most recent edition), (Cincinnati, OH: American Council of Governmental Industrial Hygienists.)
4. NFPA 45. *Fire Protection for Laboratories Using Chemicals* (Most recent edition), (Quincy, MA: National Fire Protection Association.)
5. NFPA 91. *Standard for the Installation of Blower and Exhaust Systems* (Most recent edition), (Quincy, MA: National Fire Protection Association.)
6. *Standard Number 49. Class II (Laminar Flow) Biohazard Cabinetry* (Most recent edition), (Ann Arbor, MI: National Sanitation Foundation.)

CHAPTER 11

Common Ventilation and Fume Hood Problems

G. Thomas Saunders

Few, if any, colleges and universities catalog Fume Hood 101, yet the majority of the graduates in the biological and chemical disciplines use these unsightly enclosures for years and, in effect, trust their health and safety to the correct use and operation of the hood and its associated ventilation system. There is no way to use this chapter as Fume Hood 101; however, we can certainly outline the course and leave you with viable cram notes so that you can aid in your own survival. You might even pass this information on to your colleagues and they, in turn, can be healthier survivors.

Fume hoods do not exist by themselves as they are part of a complete system: hood, exhaust air, make up air, and controls. So that we can minimize confusion and not get lost in the forest, let us consider the total subject in separate discussions and then, as the grand finale, we shall combine the areas and present our check list.

1. FUME HOOD DESIGN

The hood chamber is referred to as the hood superstructure. Reduced to its major components it has, besides the obvious sides, top, and work surface, a back baffle system, a sash(s), and hopefully a front streamlined air entrance profile.

The back baffle was devised so as to more evenly distribute the air across the face of the hood proper. This back baffle system in its proper adjustment constitutes one of the major areas of fume hood design. This baffle has two and sometimes three slots as part of its construction. Refer to Figure 11.1.

The relationship of the opening sizes of slots "A" and "C" has a major and

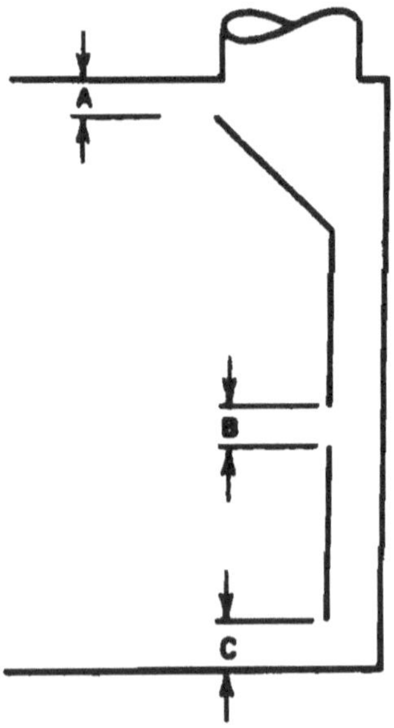

Figure 11.1. Back baffle system in a fume hood.

deciding role as to how the hood will operate. Slot "B" is a handy opening but is not a necessary part of the system.

Let us now visualize how the air behaves as it enters the hood chamber from the room proper. Consider that the hood air patterns are divided into two segments: floor sweep and vortex. The floor (or work surface) sweep is that area from the work surface to a point approximately 8 in. above the work surface. Then there is a semineutral zone from the 8 in. height to 20 in. above the work surface. The vortex is from 20 in. above the floor to the top of the hood chamber and therefore all of the area behind an open or partially opened sash.

Slot "A" (see Figure 11.1) is closest to the exhaust duct (blower) and has greater potential suction than is available to the bottom slot "C". If slot "A" is fully opened then slot "C" is at a reduced flow. As the air mass enters the hood and heads for slot "A", not all of this air can pass through the slot proper on the first pass. Some air goes past the opening, curls around, comes down behind the sash, and tries it all over again. These attempts cause "the vortex". This circular (and turbulent) air pattern can and does take the generated hood fumes and bring them from the back of the hood out to the front plane of the hood. If slot "A" is fully opened and slot "C" is closed then the vortex can extend well down into the hood chamber. To minimize the size of the vortex and vastly improve hood performance slot "A" should have an opening set at between 1/2 and 3/4". This can improve your hood performance by a factor of 1000.[1] It does not cost a cent and may save a hood-induced injury. Concurrently, slot "C" should always be fully opened (2 to 2 1/2"). If your hood has an adjustable baffle cover for slot "C", take

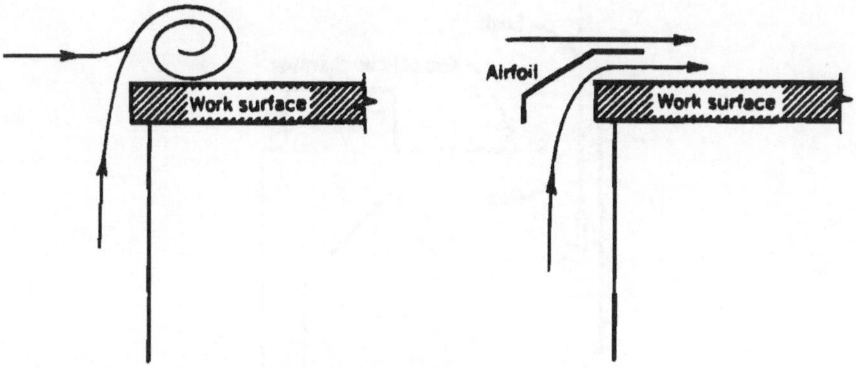

Figure 11.2. The effect of an airfoil on airflow.

it off and discard it as it can do nothing but cause problems for hood performance. An open bottom slot "C" also insures good floor sweep. Except for heat-caused density changes almost all chemical and biochemical reactions have a heavier than air density. The purpose of the hood can best be served by exhausting these fumes, so leave the bottom slot wide open. Do not believe the claims of some hood manufacturers for adjustments to accommodate heavier than and lighter than air reactions. Except for high heat loads your densities are 99% plus heavier than air, so adjust your hood baffles accordingly.

We mentioned at the beginning of this section that there were two major components in hood design. We have explored the back baffle, now we shall explore the second area: the front entrance shapes.

The vast majority of the hoods manufactured since the early 1960s have a streamlined front entrance profile. It is referred to as the air foiled shape and has a tapered configuration on all four sides.

The two side wall foils add to hood performance, the top front foil only slightly, but the bottom air foil at the work surface is a critical part for good hood performance. Air is lazy and it takes a very predictable path in going from point to point. Air entering the hood at the work surface not only comes from directly in front of the hood face but also up from the area in front of the hood base cabinets. The vectors of the vertical and horizontal air components meet at the front edge of the work surface and result in a very turbulent area. This turbulence can actually reach back into the hood and pull generated fumes to the front edge. Now you perform magic and add the front air foil; the turbulence disappears instantly (Figure 11.2). Regardless of the style of hood that you have (with the obvious exception of walk-in hoods) add an air foil if you do not have one. The cost is insignificant compared to the improved performance. Remember, leave at least a 1 in. space between the work surface and the bottom of the air foil. Otherwise your addition is self defeating and actually extends the front edge turbulence out into the laboratory proper.

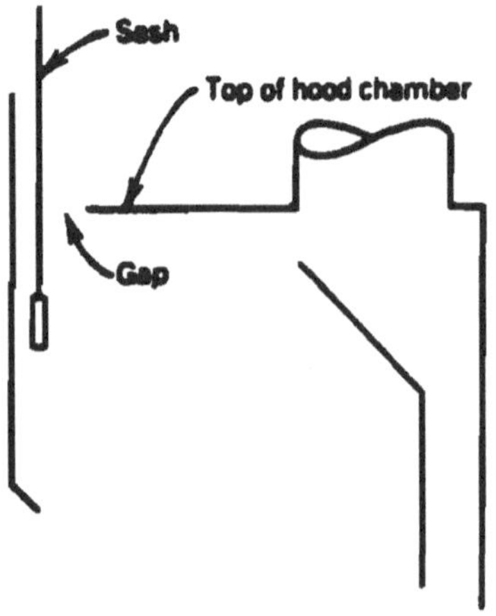

Figure 11.3. Gap at the top of typical fume hood adversely affects face velocity.

To test these two areas of design is quite easy. A 30-sec smoke bomb[2] released in the hood chamber vividly shows the vortex. A pie or small cake pan filled with warm water and some dry ice chips show floor sweep when placed about 3 to 6" back into the hood. Both tests are simple and should be part of your safety section's evaluation of hoods.

In the process of establishing basic hood construction manufacturers leave a space between the vertical rising sash and the top front edge of the hood superstructure (chamber). This is to allow the sash to go up and down without banging the hood proper. This gap, usually 1/2 to 1" allows air to by-pass the hood frontal opening when the sash is raised. This causes approximately 10% — or more depending on gap size — of the air to be wasted as far as face velocity is concerned. This is easy and inexpensive to correct. This gap should be closed with a flexible Teflon wiper (Figures 11.3 and 11.4). You should insist that it is in place when you buy new hoods and for $50-100 you can retrofit any existing vertical rising sash hood.

One last area of hood design deals exclusively with the hood front sash(s).

A common error on users' and engineers' parts is that a by-pass hood is a constant face velocity hood. A quick explanation of a by-pass hood is that this design has an open area between the top of the closed sash and superstructure proper; this area is normally covered by a grill or hidden from view by some sort of panel arrangement. A by-pass hood is a constant volume hood. The face velocity will increase as the sash is lowered and this factor varies by hood models; it ranges from an increase multiplier of from 3 to 4 1/4.

When by-pass hoods are fitted with a horizontal sash, or a combination vertical rising sash with horizontal panels, there must be some modification made to the by-pass area. The actual modification is simple. Using the same material as the hood lining you close the by-pass opening with a perforated panel with 1/4" holes

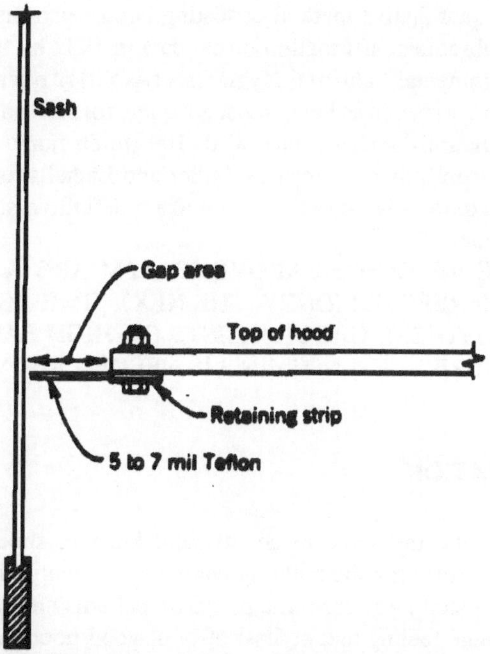

Figure 11.4. Illustration of how a Teflon wiper can be used to close gap on an existing fume hood.

drilled or punched on 1 in. centers both vertically and horizontally. When the horizontal sash(s) are open this plate makes approximately 90% of the air go through the open sash area. As these horizontal sash(s) are closed it keeps the face velocity from increasing by more than a factor of three or so.

In general, unless the hood is equipped with special face velocity controllers (variable air volume [VAV]) the hood exhaust volume is not affected when the sash is lowered. In the vast majority of existing hood systems there is no energy savings by merely lowering the sash.

To effect a savings in exhaust air (energy) you must equip the hood with a system that reduces the volume as the sash is lowered. These VAV systems come in various shapes and sizes and in a properly controlled system can and do save operating costs. When used, the by-pass area on vertical sash hoods must undergo some modification. You close the by-pass area down to the last 2 or 3 in. of sash travel. This small and controlled exhaust volume vents the hood with the sash closed but at a much lower level than with the sash open.

2. HOOD FACE VELOCITIES

In 1978 Caplan and Knutson[3] published an in-depth report on hood face velocities and a quantitative method of testing hood performance (safety). This report and the subsequent affirmation of the data in 1982 by the American Conference of Governmental Industrial Hygienists (ACGIH)[4] quantified face velocities of from 60 to 100 ft/min as being most adequate for safe hood operations. The new OSHA regulations[5] reaffirm this 60 to 100 ft/min range. This area of face velocity was concurrently described by Fuller and Etchells[6] of DuPont in 1979.

It is therefore quite safe and proper to make a definitive statement regarding hood face velocities.

HOOD FACE VELOCITIES ABOVE 100 FPM ARE A WASTE OF ENERGY AND THEREFORE MONEY. THE NEXT TIME YOUR SAFETY OR INDUSTRIAL HYGIENE GROUP INSISTS ON HIGH FACE VELOCITIES YOU CAN ASSURE THEM NOTHING IS NEEDED ABOVE 100 FPM.

3. HOOD LOCATION

Fume hoods, like the stove in an efficient kitchen, should be located for convenience to the user not the building contractor or ventilation engineer. They should also be located away from major traffic patterns and major air patterns.

It is my personal feeling that at least 50% of good hood performance can be contributed to proper room location. Here is the logic:

A hood with a face velocity of 100 ft/min has an air volume going into the opening at slightly over 1 mph. If the hood is next to a doorway, the door can provide a cross draft in the 5 mph range. A person walking by pushes and pulls an air volume of from 3 to 5 mph. An improperly designed and located air makeup diffuser can give a 5 to 10 mph cross draft. There is no hood design that can overcome such a handicap and an increase in face velocity can only add to the turbulence.

So keep hoods away from doorways if you can and make the ventilation engineer use low velocity perforated box type diffusers. The discharge velocity should be around 60 ft/min. Vane diffusers with discharge velocities at 500 to 1000 ft/min are just not acceptable. I cannot over-emphasize the diffuser selection and location. Why put $5000 into a fume hood to have it make unsafe by a $50 diffuser? Take a look at most of the laboratory facilities constructed in the past 10 years and you will discover that 90% have high velocity diffusers. What a waste and it is primarily caused by ignorance on the part of the ventilation design engineer. The sad part is that when the doors fail to perform safely, the design engineer is long since gone.

So much for the hood itself, now let us review the ventilation system that services the hood. We have discussed face velocities which are really the expression of the exhaust volume, now it is time to consider the replacement of this exhaust air.

4. MAKE-UP AIR

You cannot take the air out of a room until you put it in there in the first place: hoods are like rabbits, they seem to multiply overnight. When hoods are added to an exhaust system the blowers are adjusted to accommodate more volume but the poor make-up air system is rarely touched so as to supply more air. The common indicator is that doors become hard to open or close depending on which way they swing. If you cannot squeeze one more ounce of air out of the make-up air system, then the solution lies in establishing some type of equipment or user discipline.

Many central type systems are so constructed that all hoods are on (operating) when the system is turned on, be this 5 each 10-h day or 7 each 24-h day. In research and teaching facilities hoods are rarely in constant use. Architectural design parameters are acceptable when the hood system is rated at a 75% diversity factor (i.e., 75% of the hoods are on at one time). In actual practice many companies find that required hood use (overall) can be more commonly in the 50% diversity range. If your building control system is working and you add some proper damper control to each hood you could actually have excess exhaust and supply air. Reasonable expenditures in controls can be an inexpensive cure for some poor hood (volume) performance. The variable air volume controls that are now being recognized in the marketplace can solve some air starvation problems, but remember these modules must exercise control over both the make-up air and the exhaust air to be functional.

If these measures cannot be instituted, then you go from the hood proper to the user. Can you work with an added horizontal sash or multiple sash? By cutting down on hood opening you can reduce volume. How about having stops on the vertical sash so that the normal opening is reduced by 6 in. or more? Will your technical staff allow you to modify your vertical sash hoods with horizontal sash inserts? Viable methods but they do require something that a lot of research and scholastic laboratories lack — and that is *HOOD DISCIPLINE* — discipline to follow the rules for safe operation. The research staff at DuPont are in a constant safety indoctrination program and achieve a high level of discipline. If your facilities cannot instill this discipline, forget sash stops, horizontal sash, etc., and put in the required dollars to add air conditioning tonage and boiler capacity.

5. HOOD NOISE

Hoods in themselves do not generate noise. They do act as earphones so you can hear all of the noises generated by the exhaust system. If the noise at the hood is a hissing sound, this indicates that the duct velocity is too high and you hear air velocity noise. Air becomes audible at 1800 ft/min and starts to get annoying at 2200 to 2500 ft/min. The only solution is to increase duct sizes — sometimes going from a round to a rectangle duct gives more duct cross section without taking up more ceiling space.

The prevalent noise in duct systems is a rumble. This is the result of turbulence and/or fan noise, most likely both. The blower impeller that is not balanced can really be a pain in the ear. Do you enjoy driving your car with the front wheels in an unbalanced state? No, you get them balanced, easy to do. Balancing blower wheels is just as easy for someone who knows what to do. It also prolongs the life of the blower.

Blowers that are forced to operate at full speed because of the volume load can make an ungodly whine. Select blowers in their midrange, most blower catalogs are designed to show the quiet area of the blower performance curve; see that your HVAC engineer uses it properly.

Never judge that a hood is working well just because it makes a lot of noise. Noises and vibrations can easily transfer from one metal duct to another and it is very easy to hear all sorts of racket at a hood face when it is not even operating. Good hood systems will give you effective containment without the noises generated by excessive air velocities, turbulent turns, or unbalanced blower wheels.

6. USER DISCIPLINE

We discussed discipline briefly in regards to exhaust volume as determined by the sash configurations. Now let us consider the discipline needed to really make your hoods perform safely.

The first rule is to keep the hood as free as possible from being used as a storage area for unnecessary equipment. If you are not anticipating using something, put it in a box, a drawer, a cabinet, but not in the hood. If you are finished using something in the hood, put it back from whence it came.

Do not overload a hood with too much apparatus. It is sometimes possible to locate large items outside of the hood and run wires, tubes, or the like to the process or procedure.

When you do put equipment in the hood, keep it up off the work surface at least 1 in., preferably 2 in. Use large rubber stoppers, baby food jars, pieces of pvc pipe, or metal tubing, but use something compatible with what you are doing in the hood. Do not block off the bottom back baffle slot with cans or bottles or boxes. Just imagine that you are the cloud in the T.V. tissue comercials and you need an easy path to go into and through the hood chamber. You can solve a lot of hood problems by a proper equipment loading.

The DuPont team of Fuller and Etchells[6] points out in their research papers that putting your work at least 6" back into the hood chamber greatly increases hood safety.

7. SYSTEM MAINTENANCE

If you do not maintain your system, it will eventually fail. Insist on periodic inspections of the blowers, ducts, and controls. Do not let the maintenance people by-pass or disconnect an electronic part because they do not know how to fix it. There must be a supply of replacement parts for instant access; most hood systems cannot afford to be down for hours or days.

8. SYSTEM INDICATORS

The hood user must be able to determine that his or her hood is indeed working. The new OSHA regulations[5] will demand that such a monitor be in place and operating.

This is not a requirement dictating a complex electronic monitor; it can just as easily be an indicating manometer connected to the hood's exhaust duct stub. But whatever is used, the operator must be instructed in what it does and how it detects hood operation; then the user must have the discipline so as to refer to this device prior to going into the hood with a process or procedure.

9. CHECK LIST

We have covered the broad subject of fume hood systems but not with overwhelming detail. The main topics we covered should be placed in a priority list so that you can check your system, evaluate your overall hood performance, and make, or at least suggest, those steps that can make your facilities a safer place in which to work. This priority listing is characterized by evaluating the parameters that must be in place to give optimum hood performance. I do hope that I have given you adequate information to sustain this list and that hopefully you will agree with me.

1. Room air patterns
2. Hood location
3. Hood design
4. Adequate face velocity
5. User discipline

If you start at the top of the list and make corrections your hood performance will improve with each item. Start at the bottom and you accomplish little until you have reached item one.

References

1. Knutson, G. W., "Effect of Slot Position on Laboratory Fume Hood Performance", *Heating, Piping Air Conditioning*, 93 (Feb. 1984); Saunders, G. T., "A No-Cost Method of Improving Fume Hood Performance," *Am. Lab.* 102 (June 1984).

2. E. Vernon Hill Co., Corte Madera, CA.
3. Caplan, K. J. and G. W. Knutson, "Laboratory Fume Hoods, A Performance Test," RP70 *ASHRAE Trans.*, 84, (I)(1978).
4. Industrial Ventilation Manual, 17th ed., American Conference of Governmental Industrial Hygienists, P. O. Box 1937, Cincinnati, OH.
5. OSHA Regulations 29 CFR Part 1910, 1450.
6. Fuller, F. H. and A. W. Etchells, "The Rating of Laboratory Hood Performance," *ASHRAE J.*, (1979); p. 49—53; Mikell, W. G. and L. R. Hobbs, *J. Chem. Educ.*, 58:A132, (1981).

CHAPTER 12

Basic Concepts for Improving Ventilation During Major Remodeling Projects

Robert Kee

Major remodeling projects frequently are initiated because most older laboratories were built before very serious consideration was given to ventilation requirements. The advent of OSHA, FDA, RCRA, and related federal, state, and local regulations has brought ventilation to the forefront in laboratory upgrading. However, the total laboratory design must be considered in upgrading ventilation because positioning of laboratory furniture and relationship and orientation of various required equipment pieces seriously impact ventilation requirements.

Therefore, before any upgrading project is undertaken, it is imperative that an intensive survey of present and future occupants be made to ascertain their needs and wishes. Inevitably, there will be compromises here. Experience has shown that this survey will turn up the need for more laboratory space per scientist primarily because of greatly increased use of instrumentation and computers which take up considerable space. Also, increasingly sophisticated instrumentation and research needs are more and more demanding precise temperature and humidity control.

A thorough design review will show many, if not all, of the following problems:

- Fume hoods
  Poor design resulting in poor performance

Wrong location
Inadequate size

- Insufficient or no make-up air
- Location of make-up air adversely impacting fume hood performance
- Open windows affecting fume hood performance
- Bench top chemical operations with no provision for exhaust
- Solvent storage with no provision for exhaust
- No or poor temperature/humidity control
- Zone temperature control of large areas
- No provision for temperature/humidity control for specialized instruments or processes

There are, of course, many solutions to these problems and help is available from many sources. Consultants specializing in laboratory design are available. It is also important to engage an architectural firm with experience in laboratory design and health and safety issues. This is especially important to ensure coordination of laboratory esthetics, function, serviceability, and compliance with pertinent regulations. Visiting recently completed new or upgraded major laboratory buildings should be an important early step after an analysis of problems in your laboratory. Most companies are willing and even glad to show successful projects since there is little proprietary information to be revealed.

The following are some suggested solutions to the above-mentioned problems.

1. FUME HOOD DESIGN

The fume hood is central to all laboratory design. It is here that all toxic or otherwise hazardous operations should take place. Here, most fires and explosions that occur in laboratories happen. Clearly, they are there to protect the laboratory occupant from undue exposure to hazards. It is only in the last 15 to 20 years that much thought has been given to fume hood design.

There are many things which impact fume hood operations. First and foremost is the design of the hood which is covered in another chapter in this book. Location of the hood to minimize traffic past the face is critical to ensure that undue eddy currents do not draw fumes out of the hood. The location and method of introduction of make-up air are critical. Jet-type diffusers near the hood face can ruin the performance of the best designed fume hood. Low velocity air introduced at a point remote from the hood face minimizes supply air impact on performance. Current needs of research combined with governmental requirements are increasingly demanding larger fume hoods which require larger quantities of make-up air.

2. MAKE-UP AIR

It is obviously important that air balance be maintained in the laboratory building. The basic principle of no recirculation of potentially contaminated air imposes the requirement of large quantities of make-up air. Most older buildings have very little or no make-up air, depending on open windows and infiltration to provide the required air. Wind or lack of air adversely impacts fume hood performance. Open windows should be avoided and used only in emergencies. Use of fixed sash can many times save enough to contribute to the added cost for controlled make-up air. Very little opposition by the users has been encountered, particularly when good temperature control is provided.

The use of at least a 5-min air change where chemicals are handled has proved to be a reliable starting point. The number and size of fume hoods may dictate even larger amounts of air.

With modern research requiring more temperature and humidity control, the need for conservation of air becomes apparent. The obvious place to save is the amount of air discharged. There are as many opinions on the needed face velocity as there are experts in the field. Experience of many years of some of the largest laboratories in the country has demonstrated good safety results with as low as 60 ft/min face velocity. Government regulations, at times, have required considerably higher velocities. There is considerable research going on in this field. With ventilation costs being a major factor in operating a lab it is important to establish the lowest allowable velocity early in the design phase.

Next, with larger fume hoods (10′ or larger) consideration should be given to sliding sash hoods which cut air requirements by 50%. This is an emotional issue with many scientists believing they require full face opening at all times because of several people working in the hood at the same time. This could, of course, be a factor, and no decision should be taken here without full input and cooperation of the user. The stakes are so high, however, that this should be seriously considered and in many upgrading projects space limitations may make such an approach the only viable solution.

3. BENCH TOP OPERATIONS

Bench top operations should never be undertaken using toxic materials which should always be used in a hood. However, there are many operations which involve noxious materials which are unpleasant and are frequently performed safely at laboratory benches. Many labs use flexible, corrugated ducts (elephant trunks) to remove these fumes and they should be considered to minimize need for hoods. It should be emphasized that they should never be used for toxic or otherwise hazardous materials.

If used, they should be placed as close as possible to the fume source to maximize the capture velocity. Many people have found that the use of a sheet metal funnel at the capture end of the elephant trunk is useful in picking up fumes. Each use should be carefully considered and the application designed to fit the conditions on the bench.

4. SOLVENT STORAGE

It is quite common to store solvents in the laboratory. Most old buildings have no provision for storing these in a safe place. A fire-resistant, UL approved cabinet connected to the exhaust system to maintain a negative pressure should be used.

These cabinets are frequently used without an exhaust connection. While this minimizes the possibility of a fire in the cabinet, it does not prevent the escape of flammable vapors into the laboratory. Maintaining a negative pressure in the cabinet is the best insurance for safe storage. A good rule-of-thumb for solvent storage in a laboratory is to allow the storage of minimal quantities not to exceed the daily usage.

The small amount of air used for maintaining a slight negative pressure in the cabinet normally adds an insignificant quantity of solvent fumes to the exhaust system and would not constitute a fire hazard. It would be prudent to evaluate the dilution effect in the exhaust system if large quantities of fumes are exhausted.

5. TEMPERATURE/HUMIDITY CONTROL

Most modern laboratories have some form of instrumentation, computer, or analytical equipment which requires 24-h precise temperature and/or humidity control. It is often not feasible to provide these conditions on the central system because of the continuous operation requirement which would necessitate maintaining the large system in operation for a small number of rooms. Those requirements must be determined at an early design stage to ensure their consideration in the final design.

6. REPLACEMENT AIR

All of these air requirements eventually must be accommodated by introducing air throughout the building, customarily by distribution ducts. Space for ducts is usually quite limited in old buildings. Usually, corridor ceilings can be lowered to provide space for distribution ducts with entry to the laboratory at ceiling level. Picking up air at the lowest level (ground floor) and exhausting at the roof level has usually proved to encounter the least amount of contamination of fresh air from the exhaust. The location should be away from loading docks or parking areas to preclude admission of exhaust fumes from idling motors.

The ground level location introduces the need to move the air from the lower level to upper floors in multiple story buildings. There are at least two alternatives here. One is to sacrifice space in the building to provide a mechanical equipment room and space for duct risers. The other, which has often been found to be more expedient and less costly is to build a mechanical equipment room at ground level and build a duct chase to the roof with exits at each corridor external to the building.

A different approach which has been used successfully is to put both supply and exhaust at the roof level. This has the disadvantages of vibration from large fans which are very undesirable in a laboratory environment and also increases the possibilities of recirculation of contaminated air. Detailed studies of wind and adjacent obstructions which could impact air currents are necessary to decide on an exhaust stack height. Many times, this results in the need for a very large, high stack which is esthetically undesirable.

7. ENERGY CONSERVATION

A cost analysis is now necessary to determine the feasibility of heat/cooling recovery. This can often recover as much as 70% of the initial energy costs and can be cost effective. There are numerous heat recovery devices available but if a decision has been made for ground level air entry and roof-top exhaust, the choice is limited to heat recovery coils in the exhaust and supply with interconnected piping. To make heat recovery feasible, it is necessary to collect fume hood exhaust in a common duct to minimize the number of coils and piping. One successful method employs an exhaust plenum with a controllable variable speed fan which maintains a fixed static pressure in the plenum. The individual hoods employ a constant-velocity damper controlled by an on-off switch at the fume hood.

With the roof top installation having adjacent supply and exhaust, there are a number of different types of recovery devices available. Heat wheels and flat plate air to air collectors are two which have been used successfully.

8. TEMPERATURE CONTROL

Most older buildings that have any kind of temperature control have large zones controlled by one or several averaging thermostats. Early consideration of temperature control needs may make it possible or even necessary to provide individual room control. This is, of course, the best solution for the comfort and needs of laboratory occupants with widely varying heat loads. Good design might very well provide this at little additional cost.

In summary, the initial input of occupant's needs is probably the most important and most frequently overlooked phase of upgrading. If the project is large enough, a mock-up laboratory demonstrating the proposed design can be the key to widespread acceptance and understanding of the design. There are, of course, no standard answers to any of the above problems. Each old building is a separate case with the final solution being dictated many times by the structural limitations of the building. However, innovative design can almost always provide a modern, upgraded laboratory in any structurally sound building at considerably lower cost than dismantling and building new.

An open mind on innovative solutions that may be unconventional but functional is necessary for a successful upgrading.

Section IV: Putting Principles into Practice

CHAPTER 13

Safe Laboratory Design in the Small Business

Lyle H. Phifer

Sometime back I had occasion to visit a local paint formulator's plant specializing in production of specialty automobile paints. The place was literally a garage operation with six employees, three of whom were members of the owner's family. There was not a soul in the place with any chemistry background. The laboratory consisted of two sections of kitchen cabinets located in what might be described as a large closet. The laboratory equipment consisted of a balance and a small old laboratory oven. Three control analyses were being run — total solids by weighing a sample into a paper cup and evaporating the solvent in the oven, viscosity by sticking a glass tube in the paint and observing the dropping rate, and color comparison by dipping metal strips into the paint and after drying in the oven, visually comparing with other strips. No safety glasses were being worn and no gloves or other protective clothing was evident. No fire extinguisher was in sight. There was no ventilation except a window fan which was off the day I was there because it was cold outside and that would be a waste of heat. Obviously, the odor of paint solvents was very strong, but nobody seemed to care. I made the comment to the owner that he would be in real trouble if an OSHA inspector showed up. His response was that he was not worried about OSHA because he had too few employees. Apparently OSHA or the Pennsylvania Labor and Industry inspectors did show up eventually, and he is now out of business.

Sad to say, this type of situation is altogether too common even in this day of at least political consciousness of employee safety. But a complete appraisal of the operation gives some clue as to why it exists. Yearly sales were probably

$100,000 to $200,000. Rent, electricity, telephone, heat, and wages meant at best a very marginal profit. The owners' philosophy was simple — don't spend money on anything which he did not feel was absolutely necessary. He even complained about the cost of his balance.

The term "small chemical business" is a very broad term — it can be a small operation like the paint producer to an operation with several hundred employees and a multi-million dollar income. Regardless of the size, there are several things they have in common.

The research laboratory usually does not exist, usually because there is little research going on. Most operations do have a control or analytical laboratory. The owner or owners usually approve of virtually all major expenditures with sometimes $20 or more being a major expenditure. It is very much a rarity when an architect is hired to design a laboratory, more often it is simply a room or area designated for this purpose. The management rarely has any experience in designing the layout other than what he has seen when visiting some other operation. Unless they have had an OSHA inspection, they do not generally have a copy of CFR 29, the labor section of the Code of Federal Regulations. Even the American Chemical Society is not necessarily a help, since estimates are that less than 15% of chemists or management associated with small businesses are members of the Society. In summary, there is a remarkable ignorance on the part of small business as to what really constitutes a safe laboratory, and in the cases where some knowledge is available, there is a strong tendency to go only so far as absolutely necessary.

Let's take a look at what can be considered to be absolutely necessary in the basic design or layout of a laboratory:

1. Two exits with doors opening outward preferably into an offset instead of directly into a hall. Exits should be at opposite ends of the laboratory and easily accessible from all locations.
2. Aisles at least 4 ft wide, but if they are too wide they become storage areas.
3. Sufficient chemical fume hoods for any employee working with hazardous chemicals to be able to use one without crowding. The hoods should be located in such a position that they are not opposite desks or areas in which employees might congregate.
4. Desk locations should be carefully selected so that spills or other accidents do not endanger an employee sitting at the desk.
5. Safety showers and eye wash fountains must be at easily accessible locations for all employees working in the laboratory. The location should be at such a place that employees would not be tempted to place boxes or other items in such a position to restrict access to the shower or eye wash.
6. Sufficient grounded electrical receptacles that no extension cords are necessary and no cords will be on the floor or across aisles.
7. Proper fire extinguishers located preferably at the exits to the laboratory.
8. Provision for any gas cylinders to be located out of the main traffic flow and properly strapped or chained.
9. Bench or table tops which are impervious to any material which might be spilled.

10. Storage space for flammables.
11. Sufficient shelving or cabinets that no bottles need to be stored on the floor.
12. Adequate lighting.

All of these items can be considered to be structural in nature. One can have a safe layout but still have the least safe operation. It is really development of a set of safety rules and procedures that defines how safe the laboratory is. One obviously should start with the OSHA rules, but these are not really adequate to cover anything but the most obvious situations.

The American Chemical Society's Committee on Chemical Safety has published a safety manual for small chemical businesses.[1] This publication, which had input from numerous members of the committee, the Division of Chemical Health and Safety, and the Division of Small Chemical Business, provides a set of safety guidelines for management as well as for employees. Safety in the laboratory very much depends on careful consideration of the consequences of doing anything. Hopefully the use of the knowledge by the many people who contributed to this manual will be reflected by a safer small business laboratory with fewer accidents and a safe place to work.

Reference

1. *Chemical Safety Manual for Small Businesses,* (Washington, D.C.: American Chemical Society, 1989). This manual can be ordered by calling 800/227-5558.

CHAPTER 14

Safety Considerations in the Renovation of High School Science Laboratories

John Severns

In May of 1989, a $12 million renovation of the Urbana, Illinois, High School was completed. The project budget included extensive remodeling of 177,000 ft^2 and a 75,000 ft^2 addition to suit the needs of a 4-year high school program in contrast to the previous 3-year program. This remodeling included major renovation of the interior to reflect functional, code and maintenance criteria; replacement of the heating plant; a new ventilation system; replacement and/or upgrading of the plumbing, power, and lighting systems; and air conditioning of selected areas.

This chapter will focus on the safety considerations used in the relocation of the science laboratories, which were a part of this project. Although the remodeling of the science laboratories was not a major factor in determining whether to undertake the renovation, the fact that the project was undertaken presented an excellent opportunity to address safety considerations in these facilities.

1. PROJECT OVERVIEW

Urbana has a population of approximately 35,000. It is served by a single high school. Upon completion of the renovation, the four-year high school will accommodate 1200 students.

The original high school was constructed in 1914 and this unit housed the science laboratories and support facilities. Subsequent additions were constructed in 1925, 1955, and 1964. There were 13 separate floor levels in the complex which was from 2 to 4 stories in height.

The concern for life safety in the science laboratories was viewed as part of the total life safety planning for the project. It was decided to relocate the science laboratories to "found" space in the 1925 addition which had been used for a swimming pool and gymnasium. The space reassignment allowed consolidation of the Science Department as well as consolidation of the other departments and facilitated phasing of the construction process. Construction of the labs in this "found" space made it possible to consolidate utility services, ventilation systems, and hood exhausts in contrast to the original lab locations in two different areas of the building.

The science laboratories and library were relocated in space that had been the swimming pool on the first floor and gymnasium on the second floor. Three levels were created by adding a third floor in the 2-story gymnasium space. The space was allocated as follows:

First floor

a. two pairs of biology laboratories, each pair with a preparation room
b. science storeroom

Second floor

a. two chemistry laboratories and a preparation room
b. two laboratories for physics, general science and earth science and a preparation room
c. faculty office/workroom

Third floor Learning Resource Center

a. library
b. audio-visual services
c. one classroom

2. CODE CONSIDERATIONS

The State of Illinois enacted a school safety code entitled "Efficient and Adequate Standards for the Construction of Schools", Circular Series A, No. 156, which is the operative code for new school construction and additions after July 1, 1965. A companion code for renovation, Circular Series A, No. 157, applies to renovation of schools constructed prior to 1965. These codes take precedence over local codes and, in conjunction with the regulation of the State Fire Marshal, guide school construction and renovation in Illinois. It should be noted that School Districts in Illinois are creatures of the State, separate from municipalities, and are separate taxing bodies.

Although all of the Urbana High School was completed prior to 1965, it was the consensus of the Urbana Board of Education, District 116, the State Board of Education architect, and the architect/engineer that the A156 new school code be used due to the extent of remodeling, amount of "found space", and size of the addition.

Reference is made in A156 to the use of listed accepted standards "to be followed in situations not specifically covered herein and where specifically referred to:". These include National Fire Protection Association (NFPA) 101, Life Safety Code (1967), and the Basic Building Code, 1965, with 1966 & 1967 supplement. Building Officials Conference of America (BOCA) is cited as one of the model codes to be used for cited exceptions. Handicap Accessibility is also mandated by A156. The City of Urbana utilized the BOCA Code, 1984, Edition. These codes, and later editions of the Life Safety & BOCA codes were used to guide the design of the renovation and addition.

The School Safety Codes A156 & A157 are unique to the State of Illinois. They have proven to be an effective means of upgrading the life safety in Illinois schools in a cost effective manner for many years. Combined with NFPA 101 and the BOCA Code, or a similar model code, these codes provide excellent guidance to the architect/engineer (A/E) in planning a major renovation. Since NFPA 101 and the BOCA Code were also used, a generic approach would be little different other than a few specific guidelines.

Our design philosophy views code requirements as a vital element which must be incorporated in the design solution. Further, the codes often present a minimum acceptable level. In addition, there was a close working relationship with the high school administration and staff as well as review with the local code officials and the fire department. A construction manager was employed by the School Board to oversee the construction process. Early discussions during the design phase included all of the above groups. The life safety considerations of the completed project and during the phased construction had a significant impact on design decisions. Had this not been done in the design phase, the high school could not have maintained classes throughout the phased construction process over a two and one half year period.

Some of the specific code requirements from A156 are listed in Appendix A.

3. DESCRIPTION OF WORK PERFORMED

The size of the building was such that it was necessary to subdivide it into fire zones by rated fire separation in accordance with the code. The entire building is of noncombustible construction with the exception of limited areas which have timber roof systems. Automatic sprinklers were placed in these timber roof systems. Exit requirements throughout the building exceed minimum requirements. As part of the renovation, all thirteen levels of the building were made handicap accessible.

4. EXISTING LABORATORIES

It was determined in the design phase to reuse existing laboratory casework, tops, and utility service fittings which were in satisfactory condition. This was a

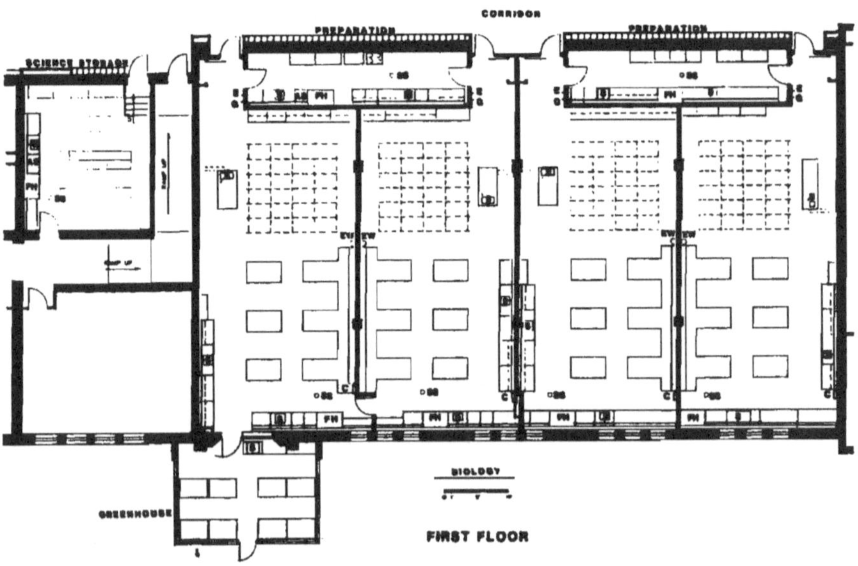

Figure 14.1. Renovation plans for biology labs and science storage.

cost savings measure. While there were measurable savings, storage and moving costs reduced the projected savings.

The following summarizes the reused and abandoned equipment and services:

1. About 60% of the existing laboratory casework was reused. The usable lab equipment was refurbished and refinished for one of the chemistry labs, both labs used for physics, general science and earth science, the four preparation rooms, and store room.
2. Lab sinks and countertops were salvaged and reused wherever possible. Some tops containing asbestos were reused intact with no cutting or refabrication, this being consistent with the non-friable provisions of the legislation.
3. Existing service utilities and fittings.

    a. All existing fume hoods were abandoned because they were in poor condition and did not meet current standards.
    b. All water service fittings and electrical outlets and wiring were replaced.
    c. Portable deionized water units were abandoned.
    d. Gas fittings were reused.

4. Movable equipment was reused wherever feasible.

5. NEW LABORATORIES

The new laboratory arrangement for each floor is shown on Figures 14.1 and 14.2. Each laboratory is divided into two functional areas:

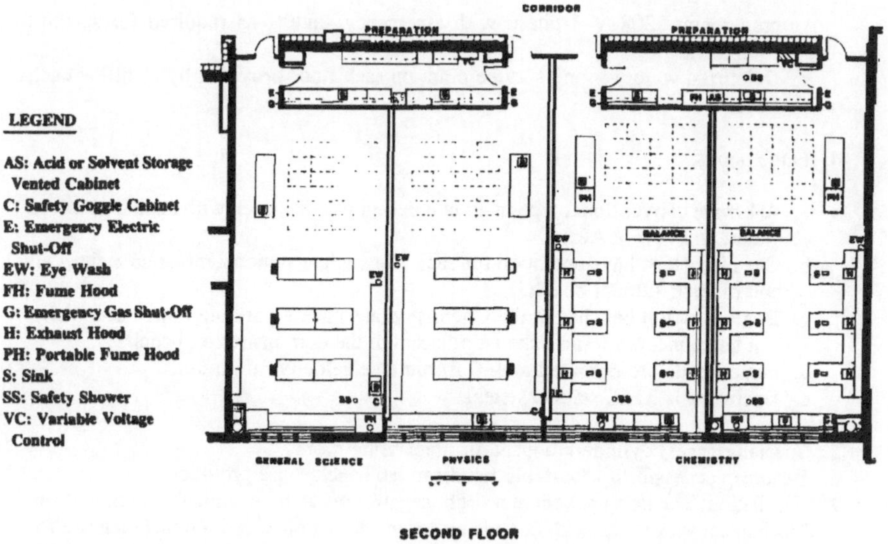

Figure 14.2. Renovation plans for general science, physics, and chemistry labs.

1. A classroom with movable tablet arm chairs and demonstration bench at the front of the room (nearest the door)
2. A laboratory area at the rear with island or peninsula benches/tables. Auxiliary laboratory equipment is also located in this area along the rear and side walls

Each laboratory is designed for 24 typical student stations. A preparation room, adjacent to the corridor, serves each pair of laboratories. A greenhouse, at grade, adjoins one biology laboratory. A science storeroom is provided on the first floor. Science offices are provided on the second floor. All departmental space is fully handicap accessible. A new elevator and ramps which make 13 different floor levels accessible is across the corridor from the science suite.

For the new laboratories, the instructional laboratories had a net assignable area of 1200 ft^2 nominal. They were designed to accommodate 24 students even though the code allows a capacity of 40. Codes require one exit. An emergency window exit is provided in accordance with NFPA 101.

The new laboratories, preparation rooms, and storeroom include the following features:

1. Casework: new and/or refurbished.
2. Tops (new): Epoxy resin.
3. Utility services:

   a. Hot and cold water with vacuum breakers.
   b. Natural gas piped to all service points. Exposed piping with emergency shutoff.
   c. 110 V AC electrical to all service points. Variable voltage to service points in chemistry, general science, and physics labs. Control panels are in the adjacent

prep rooms. 208 V 3-phase with emergency shutoff as required for specific equipment.
d. Deionized water — one service point on each floor provided by a still on each floor.

4. Fume hoods:

    a. A 4′ wide conventional type in all labs except physics; each with natural gas, cold water, and 110 V AC.
    b. Chemistry labs have one hood for each four student stations, mounted at the wall end of each student bench.
    c. Demonstration benches in the Chemistry labs have a portable fume hood which can be connected to an exhaust adjacent to the demonstration bench.
    d. Fume hoods are operated at 100 ft/min face velocity minimum.
    e. Ductwork is 316L stainless steel.

5. Specialty gases: cylinders housed in preparation rooms.
6. Handicap accessibility: portable handicap lab benches are provided.
7. Ventilation: The labs are kept at a slight negative pressure to control odor migration. Mechanical ventilation with outside air for makeup is provided. Fume hoods are for each pair of labs and preparation room. This was an economy measure and simplified installation in the existing building.

In addition to the above, the preparation rooms and storeroom each have a vented solvent storage cabinet and acid storage cabinet. Anticipated quantities of hazardous materials, liquids and chemicals will not exceed limits in Table 306.2.1, BOCA, 1984.

It would be preferable to have individual exhaust, controls, and makeup air for each hood. Fume hood exhaust fans are roof mounted and venturi discharges at 4000 ft/min exit velocity are 7 ft above the roof.

Miscellaneous safety features in the new laboratories include the following:

a. Eye washer in all laboratories
b. Safety showers adjacent to all fume hoods.
c. Safety goggles stored in a cabinet with an ultraviolet light source in all laboratories.
d. Smoke detectors tied into fire alarm system are provided in each room.
e. Exit window in all laboratories.
f. The fume hoods are located in the extreme end of the lab. Where feasible, there is a passage to an adjacent laboratory.
g. Peninsula benches have no more than two students per side.

6. SUMMARY

A major renovation project for the Urbana High School provided an opportunity to incorporate a number of safety features into the science laboratories. The existing life safety codes provided excellent guidance in most cases. Our experience suggests that the best results are obtained when the codes are viewed as minimum acceptable standards and when all interested parties are involved early in the design stage. It is hoped that our experiences and decisions can prove helpful to others.

APPENDIX A

Specific Code Considerations

Based on Circular Series A No. 156, *Efficient and Adequate Standards for the Construction of Schools*. State of Illinois, State Board of Education, July 1, 1969 and Revisions, November 1, 1974.

Fire resistive ratings of Structural Elements in 1 h (1B - h) Table A.

1. Bearing walls and their support	1 h
2. Nonbearing walls	noncombustible
3. Floors & floor supports	1 h
First and below	1 h
Other	noncombustible
4. Columns supporting roofs & roofs less than 20 feet to lowest member	3/4 h
Roofs 20 feet or more to lowest member	noncombustible

Height and area limits

1. 1 story	no limit
2. 2 story	50,000 ft²
3. 3 story	35,000 ft²
4. over 3 story	not permitted

Occupancy	**Fire resistive rating**
1. Science laboratories	3/4 h
2. Corridor partitions	not less than 1/2 h

Partitions requiring a 1/2 h fire resistive rating and doors therein may have fixed panels of 1/4" wired glass without size limitation.

3. Doors shall have closers, no hold opens.

Exits

1. Laboratories shall have at least 30 net ft² per occupant.
2. 1 exit to corridor for room with less than 1200 net ft² and not over 60 occupancy.
3. Emergency window exit in accord with NFPA 101(81) paragraph 11-2.11.5.

Interior finish	**Flamespread wall and ceiling**
1. Laboratories less than 1000 NSF	200
2. All others	75

Automatic sprinklers: not required.

Storage: required.

Automatic fire detection/alarm: Required in accord with NFPA 72A, B, C, & D. Provide audible and visual signals.

Laboratories producing objectionable fumes

1. All air exhausted from labs (including fume hoods & ceiling hoods) must total at least 1.25 cfm/ft^2 of floor area.
2. Animal rooms — 2.0 cfm/ft^2

Outdoor make-up air must equal 33.3% of exhaust up to design room temperature, then not less than 75% thereafter.

Laboratory fume hoods: Exhausts may be individual or in manifold at not less than 150 cfm/ft of length of hood opening. This equates to 100 fpm at a typical partial opening of 18 in. Make-up air may be drawn from the room.

CHAPTER 15

The Los Angeles Unified School District — A Study in the Evolution of Facilities for Teaching High School Chemistry

Gerald J. Garner

In the last 30 years, the Los Angeles Unified School District has undergone an evolutionary process in the design of science laboratory classrooms. The presentation will include a discussion of past, present, and future district high school facilities for teaching chemistry.

It has long been acknowledged that student laboratory investigation is an integral part of secondary school chemical education. Science educators agree that a complete program must strive for a balance between the learning of factual content and the development of an understanding of the process and skills of science. The processes and skills can best be learned through laboratory investigation. Laboratory activities have the potential for also assisting the learner in the development of positive attitudes about science and higher levels of cognition and thinking processes. Computer simulation of laboratory activities, although gaining in popularity, can never replace meaningful, content related, hands-on laboratory experience, particularly when the full range of possible educational outcomes is considered.

For these reasons, secondary school student access to science laboratory facilities remains a high priority. Of these facilities, the chemistry laboratory represents the most complex, presenting the greatest challenges to insure usefulness and safety.

Through the 1950s, many facilities for the teaching of high school chemistry were found with separate classrooms and laboratories. The laboratories were frequently shared by more than one teacher. This system was inconvenient, primarily because of scheduling problems and the lack of assistants to prepare, distribute, and remove supplies and equipment. Laboratories consisted of three traditional two-sided laboratory benches with reagent racks above, utilities directed toward students or a central trough, and lockers below the counters. The benches were typically arranged perpendicular to the long axis of the room. Stockrooms were adjacent and tended to be large, with ample storage.

In the late 1950s and early 1960s, a substantial gain in enrollment and the quantity of science courses offered made it necessary to construct many new facilities. Recognizing the limitations of the separate chemistry laboratory, a design was developed which used a long room with a demonstration table at one end, an area for student tablet armchairs, and three traditional-style laboratory benches in the back. It was typical to construct two of these rooms as mirror images, back-to-back, with a small shared stockroom between. It rapidly became apparent that although this design overcame the scheduling problem and permitted laboratory activity at any time convenient to the teacher, other problems were created. Because the teacher's desk and student laboratory work area were at opposite ends of the room, supervision of student activity in the last row or two of the laboratory area was very difficult. Also, because of the restricted area for student seating, the front row of chairs had to be too close to the demonstration table and the back row was against the front of the first laboratory bench which restricted access to the working area and student lockers.

In the mid-1960s, another design was developed which became the District standard until this year (Figure 15.1). This design incorporates between seven and nine laboratory stations on the periphery of the room. These are trapezoidal, accommodating two students on each side and, if necessary, an additional student at the end. The tops are flat, interrupted only with a sink and water and gas outlets. Electrical outlets are located on the ends. Student lockers are located in two rows on each side. These rooms have been typically constructed in pairs with a small stockroom between. Many classrooms have been retrofitted to this configuration as it offers the advantage of locating utility lines against the walls. These rooms were originally constructed with a demonstration table but this was replaced with an additional student module with the rationale that it could also be used as a demonstration area. This design offers complete flexibility in the scheduling of laboratory activities. The major component in secondary school chemistry laboratory safety is supervision, and the teacher can readily observe student laboratory activities from various places in the room.

Around 1970, a new building at an existing school site incorporated an experimental centralized science stockroom. This was a large interior room opening onto several laboratory/classrooms. The purpose was to determine if this type of facility improved the preparation of laboratory materials and encouraged a more efficient use of available supplies and equipment.

In 1988, committees were convened to meet with architects and reexamine secondary school facilities design. Several new considerations needed to be

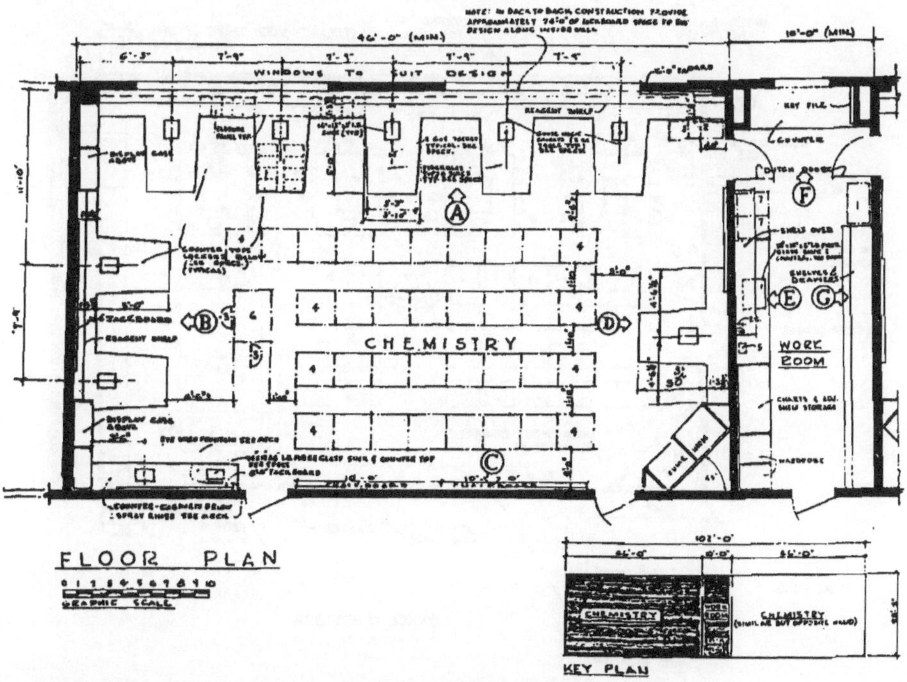

Figure 15.1. 1960s chemistry laboratory floor plan.

addressed. In the past, the available funds for school construction had permitted somewhat larger, more elaborate facilities. It was anticipated that new funding would be only from state sources and would require restrictions on available square footage. The issue of the inherent inefficiency of small, decentralized stockrooms was reconsidered. Another item for consideration was the anticipation of a major increase in the availability of software and microcomputer-based laboratories with transducer "probeware" which would permit more interactive laboratory investigations involving the use of computers.

The proposed design (Figure 15.2) for all secondary science laboratory/classrooms, including chemistry, calls for the placement of five to six island-type student work stations across one long side and one end of the laboratory/classroom, forming an "L". These work stations are 5 ft^2 and will be designed for two students to be working on each of three sides. The side facing the opposite long side of the room, or "front", is not intended for student use. The work stations will be separated from the wall by an aisle. Each work station will include a rectangular sink, water, gas, and electrical outlets. Drawers for storage will be on three sides. The wall will have a counter interrupted by spaces wide enough for two chairs back-to-back. Additional storage will be above and below the counter and in the end of the room opposite from student work stations. The counter is intended for computer locations and will have conduits extending under the aisle

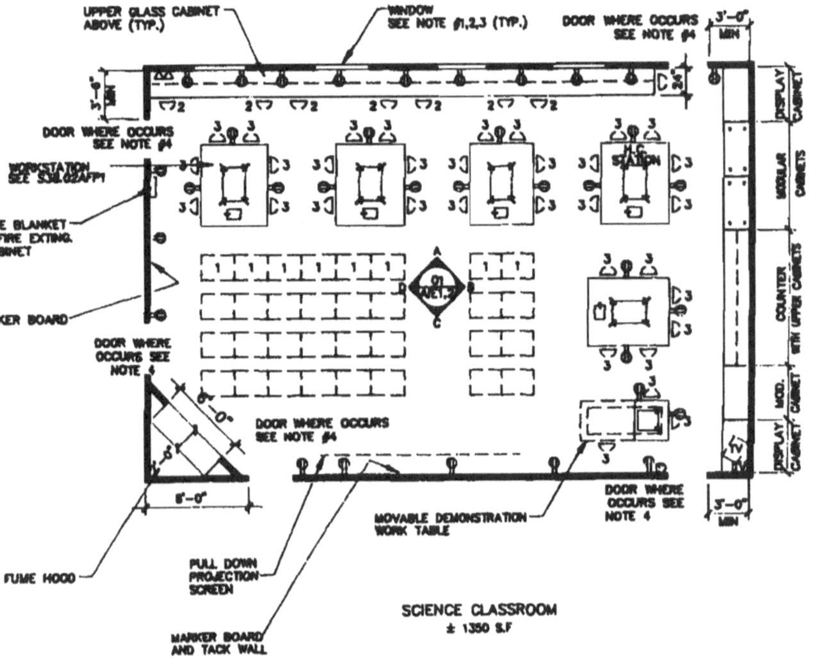

Figure 15.2. 1988 chemistry laboratory floor plan.

floor to the student work stations for the connection of probeware to be used in laboratory investigations. The front or side of the room opposite student work stations will have a demonstration table. Student seating will be at table armchairs or tables in the area inside the "L" and in front of the demonstration table. The individual rooms will not save space as compared to previous designs but a major space saving will come from the incorporation of a centralized stockroom, opening to most of the science laboratory/classrooms.

This latest design, anticipated to be the District standard for the 1990s and beyond, possesses several major advantages. Flexibility in the scheduling of science laboratory activities and an excellent configuration for teacher supervision are maintained; the importance of computers in instruction is acknowledged and a location for them is provided which is convenient, yet removed from possible contact with water, chemicals, or solutions; and the efficiencies inherent in multiple teacher use of a centralized stockroom are possible while creating an excellent rationale for the availability of an aide or stock clerk to prepare materials and manage the facility.

Laboratory/classrooms must be designed to permit maximum storage of supplies and equipment. Stockrooms must be sufficiently large to contain the necessary variety and quantity of chemicals and other supplies, to restrict storage of chemicals to no more than 5 ft from the floor, to permit inventory and rotation of stock, to provide space for cabinets for the storage of flammable materials and acids, and to provide sufficient space for preparation. Where earthquakes must be considered as a potential threat, barriers must be provided on all shelves which contain materials subject to breakage, and all free-standing cabinets and tall equipment must be securely fastened to the walls.

Countertop materials have also evolved through the years. Most chemistry laboratory countertops were originally an acid-treated hardwood, finished with a chemical resistant coating, and heavily waxed. In the early 1960s, this was replaced by a laminated, formed, fiberglass-based chemical-resistant top. This top, grey in color, was quite satisfactory but possessed one major drawback. Hot objects placed on it scorched the surface badly, leaving blackened areas. For this reason, the new standard will incorporate thick, molded, black epoxy, or similar countertops. It is anticipated that they will better withstand the chemicals normally used in high school chemistry and even if scorched by hot objects, will be less likely to show the damage.

Laboratory/classrooms must contain the facilities needed for disposal of wastes. Plumbing must be adequate. Secondary school chemistry is in great need of guidelines regarding disposal of chemicals commonly used or produced in laboratory activities. These guidelines should include identification of chemicals and quantities recommended for on-site disposal, dilution factors, procedures for neutralization, and disposal techniques. Chemicals which should never be disposed of on-site should also be identified and recommendations given on arranging for disposal.

Schools should be encouraged to recycle usable surplus chemicals within their own department and by sharing with other nearby schools, perhaps through periodic "swap-meets."

Schools and districts considering retrofitting existing nonscience or substandard science classrooms should take into consideration the potential subject-field use of a room, proximity to other science classrooms and stockrooms, and the trade-off between bringing many rooms to a minimal standard versus a few rooms to an optimum standard. One would hope that whatever the decision, that it not be based on public relations grounds but on what is best for students, teachers, and the instructional program.

California has no law governing class size in activity-based subjects other than vocational education. It is clear to those in science education that, although studies have shown that student learning is independent of class size, student laboratory activities are an integral part of a balanced science program and that student activities are not independent of other concerns such as materials management and safety. Research is needed on the relationship between laboratory-based learning, facilities design, class size, effective management, and safety. If, as many of us suspect, it is shown that class size is an important factor, this could become a basis for the drafting of model legislation designed to limit class size to optimize student achievement in science.

The optimum condition is for all students to experience a meaningful, personally relevant education in science which includes laboratory activities. Not all students take chemistry. As concerns increase regarding the state of general scientific and technological literacy, it is incumbent on all of us to insist on a quality science education program for all students. Each science teacher needs a laboratory/classroom on a daily basis, for the entire day. As local resources are not adequate, one must hope that state and federal elected officials and agencies get the message and provide the funding to give each student the opportunities which he or she deserves. This cannot help but also improve the quality of secondary school chemical education.

CHAPTER 16

Design of Safe R & D Labs

Janet Baum

1. INTRODUCTION

This chapter will focus upon the design of safe team laboratories. Team laboratories are popular in industrial research and development facilities for their flexibility, ability to promote communication, and lower construction cost. For these same reasons, team labs are gaining popularity in college and university research facilities as well. Although this discussion will focus on new facilities, key issues apply to renovation of smaller single laboratories as well.

The primary characteristic of a team lab is its size. A team or "open" laboratory is larger than typical 2 or 4 person lab units of 200 to 600 net usable ft^2. (Net usable area is the measure of the room area, in units of square feet, within the enclosing walls. Net usable area does not include the area occupied by structural elements, walls, mechanical shafts, or public circulation.) A typical team lab is 800 net ft^2 or more and accommodates 6 or more researchers. A team lab is organized on the basis of modules. The team laboratory usually occupies several or many contiguous modules. Modules are defined in two ways, by area and by function. A laboratory module is a rectangular unit of area of recommended fixed dimensions. The functional definition of a module is a space with a regular array of work stations and utilities, uninterrupted by full-height partitions, public corridors, or doorways. Functions such as offices, instrument rooms, special equipment rooms, process labs, and environmentally controlled rooms may open directly onto the the team lab, but do not necessarily conform to the modular layout of the laboratory. Module design will be discussed and illustrated in the "Design" paragraphs.

The basic design for industrial open labs has been utilized for more than a century. As the hazards of exposure to chemicals in the laboratory were recognized, measures were taken to contain contaminants, provide dilution air, and condition it for human comfort. Heating, ventilating, and air conditioning engineering design progressed to improve safety in research laboratory environments of relatively small area. Engineers could assure appropriate flow rates and constant volume of air supply and exhaust to each discrete space. In the 1970s popularity of "open office" design stimulated renewed interest in and development of large open lab environments that respond to the increasing economic constraints of building and energy costs. Team management concepts were promoted for research and highly trained workforces. Engineers again confronted the difficulties of regulating air flows, potentially containing larger quantities of varied contaminants, in larger research laboratory units. Team laboratories present unique design problems. They are more open. There are more people, movement, and activities. It is more difficult to control contaminants and contain accidents in team labs. More people may be exposed instantaneously to a chemical release, smoke, or fire. It is more difficult to control air distribution and to achieve energy conservation.

This chapter will discuss three topics on the design of safe research and development team laboratories which address these problems:

1. The stages of design and critical decisions made at each
2. Predesign and design issues for team laboratories
3. Two examples of team research laboratory designs that met the needs of the researchers and the owners

2. STAGES OF DESIGN AND CRITICAL DECISIONS TO BE MADE AT EACH

The design process can be likened to an ascending spiral. At the beginning, broad issues are investigated and decisions made on a wide range of general information. As design progresses, the issues become of narrower focus and the decisions are more and more specific. It becomes more difficult to redefine the direction or scope of the project. An experienced design team will bring forward issues for review by health and safety professionals and owner decisions in a timely fashion. Major changes that are made too late incur time and money lost in redesign.

Predesign

The first stage of a building design is predesign, which includes programming and conceptual planning activities. These initial tasks may take 6 months to 1 year for a new large lab building or extensive lab renovation. Smaller scale projects take less time. Predesign clarifies the goals, identifies users, objectives, and finances of the owner or institution. An in-house team of representatives from management, research, safety, operations, and maintenance can accomplish these tasks. Expert laboratory planning and engineering consultants can be hired to assist and direct the effort. The product of this stage of design is a written document, called a building program. A program clearly defines the performance criteria to which the new or remodeled laboratory facility will be designed. A program thoroughly describes each function which the building will house. It will list requirements for the following:

Net usable area:	For each room and function, estimate size of typical lab module
Number of persons:	Staffing pattern and population density for each room or function
Scientific processes:	Clearances, utility connections, and heat output, waste products
Equipment:	Size, clearances, weight, utility connections, and heat output
Chemicals and hazardous materials:	Quantities, reactivity, flammability, toxicity, explosivity, types of storage containers
Other hazards:	Electrical, radiation, extreme temperature, weight, laceration, infection, falls
Safety equipment:	Chemical fume hoods, special hoods, local exhaust or filtration, glasses, protective clothing, respirator, radiation shielding, fire extinguisher, sprinkler, deluge shower, eyewash, spill kit
Environment:	Temperature, relative humidity, vibration tolerance, noise level, air flow and velocities, construction materials, and finishes
Utilities:	Electricity, gas(es), vacuum, steam, hot, cold or purified water, drain, chilled or process water, exhaust, and supply air
Adjacencies:	To other functions or services, distance and frequency of trips
Materials handling:	Shipping and receiving, chemical supplies delivery to labs, waste chemical removal, waste separation, treatment, and disposal

One way to organize this data is to generate a data sheet for each room or room type in the building. Some data base computer software can format data sheets that facilitate analysis and computation of data entered. A program contains not only the raw data, but a summary of the net usable area and design and performance criteria to meet any normal and special environmental or hazardous conditions brought out in the data. The program is used by architects and engineers in the formal design process as a performance specification for each space, its area, population, contents, environment, and utilities. The program is used by owners and users as a check-list of the design documents to make sure all major factors have been addressed. There is a discussion on the programming in *Guidelines for Laboratory Design: Health and Safety Considerations*.*

Conceptual planning translates the program information into a preliminary idea of the organization of the building, or if the project is a renovation, a portion of one. Conceptual plans illustrate, at a small scale, sizes and shapes of each room, possible relationships with other rooms and the building exterior, egress routes and circulation, and mechanical and structural infrastructure.

Good programs and conceptual plans develop with ample input and review from the actual laboratory users, if they are known and participate. Programs

* DiBerardinis, *Guidelines for Laboaratory Design: Health and Safety Considerations*, p. 12-25.

should not be written or regarded simply as researchers' wish lists. With careful and thorough interviews of users, important and essential elements for good laboratory performance will emerge. Laboratory safety engineers can bring insight and valuable criticism into the programming process if they participate directly with the users to highlight safety features each lab or process will reasonably need. Cooperation in the predesign stage between users, designers, and health and safety professionals clarifies the health and safety priorities for the design. Four critical decisions at the predesign stage are the following:

1. Establish acceptable health and safety criteria for egress, fume hoods, safety and emergency equipment, fire protection
2. Estimate the current and future building population and determine the desired density to avoid overcrowding
3. Program sufficient and appropriate laboratory support spaces, such as chemical storage and high heat or high hazard areas
4. Determine the ventilation performance criteria

Design

The second stage is the design process itself. This too can take 1 year for a new building or renovation of significant size. This period should allow sufficient time for careful document and cost reviews by owners and users. The design is executed by a team of qualified architects and engineers, who are fully informed of applicable codes, standards, and regulations. They have current professional registration in the state in which the construction will take place. The design process divides into three phases: schematic design, design development, and construction documents. Generally there are meetings between the designers and owner, users, and safety professionals at regular intervals during this entire period. Good communication from owner and users to designers is essential for a successful project. A well-documented program promotes understanding of the owner's goals by the design team from the start. As mentioned before, at each phase of design, plans and specifications become more detailed. As decisions at each phase become more specific, it is more difficult, time consuming, and expensive for the owner or users to make major changes. By the end of the construction document phase, all the materials, safety equipment, code requirements, quality standards, and construction methods for each and every aspect of construction should be recorded in the drawings or specifications. A health and safety professional and knowledgable representative from the user group should carefully review the design documents at each phase. Local and state authorities also will review the construction documents for compliance to codes and standards of public health and safety when a building permit application is made. Critical decisions at the design stage are

1. Retain experienced, reputable laboratory design professionals for design services

2. Participate constructively in the design process by clearly and consistently communicating your criteria
3. Health and safety professionals review documents at each phase

Specific design issues discussed in following paragraphs are modular design and layout of team laboratories with hazard zoning, and ventilation performance criteria for team lab design.

Bidding and Negotiation

The third stage is bidding and negotiation. The nature and timing of this stage depends on the organization's financial protocols as well as the local construction climate. Bidding a major new laboratory building can take over 2 months time. Negotiation, according to whether the job was bid initally or not, can take just as long. Contractors are faced not only with attempting to get the best prices for labor and materials, but to get them on the job site at the best time. Owners should be aware of these trade-offs in bid reviews. Owner's bid forms should include the question of the contractor's commitment to a specified construction period. If a product specified in the documents can arrive at an optimal time, the job can be done more efficiently, at lower cost, even if the cost of the line item itself is greater than that in a competing bid.

The critical decisions of this stage are to not only select qualified general contractors to bid on the project but also qualified subcontractors in the mechanical and electrical trades. All must demonstrate experience in constructing a number of quality laboratories. Many organizations prefer to obtain the services of a construction manager early in the design stage to guide the owner and designers on probable costs and schedule. The group or person who fulfills this role should also have a direct and excellent track record in laboratory building construction.

Construction

Construction is the fourth stage. Construction of laboratory buildings can run from <1 to 3 years. Contractors have primary responsibility to actually construct the building and to provide day to day supervision and quality control of all materials and work done on the job site and off-site. The contractor, rather than the owner, generally applies to local and state authorities for the building permit. Local building inspectors may visit the site unannounced at any time. Inspectors can cite contractors or owners for noncompliance to codes and local regulations for construction safety on the job site as well as construction materials and methods. At scheduled times they perform inspections of rough-in plumbing, HVAC, and electrical work before the work is covered by wall and ceiling surfaces. Other scheduled official inspections take place at the completion of work of certain trades, such as sprinkler plumbing and sheetrock installation.

The design architects and engineers, within standard contractual agreements, perform regular inspections of the work to check compliance with the intent of the contract documents. Owners also may wish to hire a site representative, called a Clerk of the Works, to check daily on the labor and materials the contractor claims are on the job. Other owner representatives check that the correct materials are installed as indicated in the contract documents. Many times it is very helpful for a representative of the user group, a researcher or technician, who actually understands what the labs contain and how they are to perform, to regularly visit the site and review construction progress. Each interest group, contractor, designers, owner, and users has a different perspective on the project. Working together as a team, they benefit the construction process. Critical decisions of the construction stage are

1. Select the individual(s) who will inspect construction on behalf of owners and users.
2. Evaluate substitutions which the contractor will offer for those materials and methods specified in the design.

Project Closeout

The fifth stage is project closeout. According to the size of the project, 1 to 3 months should be set aside at the end of construction of a laboratory building or major renovation, before the users move in. This stage is often left out of project schedules and contract specifications. There are clearly defined tasks the contractor and design team must complete for project closeout. If the building is occupied, this stage may be compromised.

The design team carefully inspects the construction at the contractor's notification of substantial completion. During this inspection they produce a "punch" list of items that are missing, damaged or incorrect, to be replaced, and construction conditions which do not meet the level of quality specified in the documents. The contractor then does the following tasks: corrects all of the punch list items, tests the safety and emergency equipment and systems, performs final cleaning, certification, balancing, and start up of all mechanical systems, including adding final filters, moves and installs owner's equipment, if these were included in the contract scope. In addition, the contractor should label all pipes, valves, electrical conduits, switches, and panels according to nationally recognized standards such as Electronic Industries Association (EIA), American Society of Heating, Refrigerating and Air Conditioning Engineers (ASHRAE), and the American National Standards Institute (ANSI). ANSI publishes "Safety Color Code for Marking Physical Hazards" ANSI Z53.1-197,* and "EIA Standard Colors for Color Identification and Coding" ANSI/EIA-359-A-1984.** Furthermore, safety professionals should inspect newly constructed laboratories prior to occupancy, then again after the building is occupied to locate conditions or materials which need warning signs and labels.

* ANSI/EIA 359-A-1984, p. 1-12.

** ANSI 253.1-1976, p. ii - 14.

The local building inspector will schedule his final inspections at this time. The inspector may require written certification of all tests on safety and mechanical equipment and systems. At this stage the owner applies for the Certificate of Occupancy, or its equivalent, from the local and/or state authorities. Officially the actual move-in should not take place before this certificate, or other legal clearance, from the building department is issued. The design team and owner should investigate the local codes and ordinances for details pertaining to these legal requirements. For further information on project closeout, refer to the chapter on Performance and Final Acceptance Criteria in *Guidelines for Laboratory Design.** Critical decisions at project closeout are

1. Acceptance (or rejection) of the contractor's work
2. Schedule occupancy when the job is complete and permission is granted by legal authority

Training

Occupant and operating personnel training is the final stage of a building project. It may be done concurrently with a well-organized project closeout or afterwards. In either event, it must be done. Safety professionals who are responsible for setting and promoting safety standards in the organization should schedule and conduct walk-throughs of the project with all occupants. All occupants should understand the meaning of alarms, use the safety equipment, know the location and use of emergency shut-off systems within and outside the laboratories. They should also understand the basics of the ventilation system operation, including fume hood operation. Occupants should know all egress routes. All signs indicating egress, egress routes, and the location and proper operation of emergency and safety equipment should be permanently affixed before occupancy. Fire drills and emergency evacuation protocols should be conducted soon after occupancy.

The contractor and relevant subcontractors should conduct a complete walk-through of the project for the facilities engineers, safety officers, and equipment operators. The design engineers and architect should accompany them to familiarize building operators and maintenance personnel with the performance objectives, the equipment, and most important, the controls of the project. Particular attention should be placed on all alarm systems, safety controls, and shut-offs. It is critical that contractors draft "as-built" drawings of all materials and conditions which differed in execution from the design documents. These "as-builts" are provided in reproducible form to the owner for future reference. Contractors also assemble multiple copies of operating and maintenance manuals for the owner's operations engineers and the laboratory manager. The O and M manual contains all warranties, guarantees, operating and maintenance instructions, parts lists, and manufacturer's service protocols.

* DiBerardinis, Baum, *Guidelines for Laboratory Design: Health and Safety Considerations*, p. 193-202.

Table 16.1 Typical Activities and Materials Found in Specific Lab Types*

Lab type	Typical hazards
General	Small amounts of toxic chemicals, volatile liquids, flammable liquids, compressed gases, dusts
Biochemistry	Same as general chemistry plus microbiological agents, radioactivity, carcinogens, mutagens, teratogens, high voltage
Physics and physical chemistry	Same as general chemistry plus high voltage and current, lasers, gas pressures over 2500 psig, liquid pressures over 5000 psig
Organic and inorganic	Large amounts of toxic chemicals, volatile liquids, flammable liquids, compressed gases, dusts
Theoretical	High voltage and current
Radioisotope	Same as general chemistry plus concentrations of radioactive materials

There is only one decision prior to occupancy, that is to provide training for all occupants and building operators.

3. PREDESIGN AND DESIGN ISSUES

There are several tasks in the actual design process that have a significant impact on the success of a laboratory project. They are the following:

1. Evaluate potential hazards and the safety equipment in the laboratory required to reduce risks
2. Apply principles of hazard zoning in the layout of team laboratories
3. Plan for future expansion and avoid overcrowding

Evaluate Hazards and Safety Equipment

The building program, developed in predesign, lists hazards for each room: chemicals, materials, and other types of hazards. Health and safety professionals should study this data to make sure it is complete and accurately reflects severity of the risks. They may recommend strategies for appropriate fire protection, safety, and industrial hygiene measures. They have experience to evaluate overall liability to the users and to recommend specific safety equipment and systems to the design team. Different zones of the building will have different risks. Different laboratory types represent a variety of risks, as illustrated in Table 16.1. Definitions of and distinctions between laboratory types mentioned below as well as safety issues particular to specific types of laboratories are discussed in detail in *Guidelines for Laboratory Design.***

* Baum, DiBerardinis, American Chemical Society 196th Annual Meeting, Committee on Chemical Safety Symposium, September, 1988.

** Ibid., p. 89, 92, 105, 130, 170.

Table 16.2 Chemical Fume Hood Distribution Based on Laboratory Occupancy

Lab type	Linear feet of sash opening per research worker
General chemistry	3
Biochemistry	3—6[a]
Physics & physical chemistry	3[a]
Organic chemistry	6—12
Inorganic chemistry	5—6
Radioisotope chemistry	6—8[a]

[a] Radioisotope hoods are required where minimum concentrations of certain radioactive isotopes are used. Radioisotope hoods specially designed to facilitate decontamination.

Many of the laboratories listed above require chemical fume hoods to contain and dispose of hazardous chemical fumes. The type and distribution of chemical or special fume hoods is a very important issue both for safety and building cost. Table 16.2 lists preliminary planning guidelines on determining the fume hood density for various laboratory types. Apply the population of each laboratory, as compiled in the program document, to this guide to calculate the number of fume hoods recommended in each laboratory and for the total building.

Safe operation of chemical fume hoods depends strongly on their proper location within the laboratory. Designers should lay out laboratories so that fume hoods are located away from major egress pathways and laboratory traffic. In ASHRAE/ANSI Standard 110-1985, the American Society of Heating, Refrigerating and Air Conditioning Engineers (ASHRAE) and American National Standard Institute, Inc. (ANSI)* and others** have documented poor, unacceptable performance of chemical fume hoods when air currents near the face opening of the hood are disturbed. This can happen with improper use of the hood, equipment placed near the face opening, when people walk near an open sash, or when supply air diffusers create excessive air velocity or turbulence. Figure 16.1 shows examples of preferred fume hood locations in double module layouts.

Team laboratories, being large, may contain multiple fume hoods. Air currents and traffic around multiple fume hoods can seriously compromise fume hood performance. There are several strategies the designer can apply to these situations. Fume hoods can be arrayed together in a line along one wall. However, the access aisle in front of fume hoods should not function as a primary egress pathway. A secondary egress pathway in front of multiple hoods is acceptable. This secondary egress should lead to more than one exit. If more than two fume hoods are installed, the aisle should not be dead end (see Figure 16.1)

Another strategy is to separate the fume hoods from each other and locate each

* ANSI/ASHRAE Standard 110-1985.

** Chamberlin, "Laboratory Fume Hood Specifications and Performance Testing Requirements".

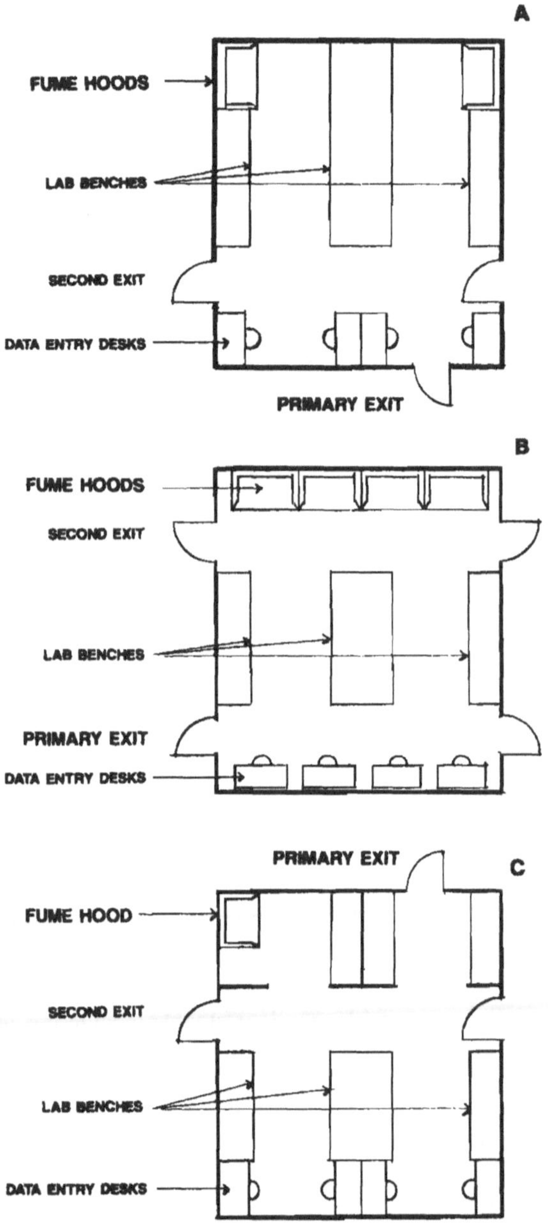

Figure 16.1. Examples of chemical fume hood locations in laboratories.

in an alcove which is near, but not directly adjacent to a primary egress pathway. The alcove provides an aisle, without through traffic, in which the air flows to the face of the fume hood can be carefully controlled. Figure 16.1 illustrates this. Fume hood layout should also be designed with the concept of hazard zoning, as explained in a later paragraph.

Health and safety professionals should consult with the design engineers to specify the types of hoods to be installed. In typical chemical fume hoods three primary types are available: standard, by-pass, and auxiliary air. Each has different air flow characteristics and energy consumption values. Current engineering practice allows some control of the volume of air supplied to the sash opening for energy savings. Visual and audible alarms can be attached to the hood to indicate low face velocity and unsafe operating conditions. Please refer to Section III in this book for more information on typical types of fume hoods and ventilation systems design. Special types of fume hoods encountered in research and development laboratories include radioisotope, perchloric acid, and "California" hoods. Materials and processes used in these hoods represent special hazards in the open team laboratory. Consideration should be given to locating them in rooms separate from team laboratories.

The interior work area of a radioisotope hood is completely lined with welded stainless steel, a material which can be decontaminated. All corners are cove shaped for easy cleaning. Welded stainless steel is recommended for exhaust ducts and risers at radioisotope hoods, because the material can be decontaminated, lasts longer, and is more durable than galvanized metals or plastic materials. According to concentrations, types, and quantities of isotopes used in these hoods, state and local regulations may require one or more of the following accessories for each hood: an isotope trap (activated carbon filter installed at the fume hood), activated carbon filter, high efficiency particulate filter, absorption filter, and scrubber (installed in the riser prior to entering the fan), and an isotope sampling pump (installed at the hood or stack). The Nuclear Regulatory Commission licenses individual research investigators and organizations for the use of radioisotopes. Refer to *Code of Federal Regulations, Title 10,* Parts 1 through 70 for more information.* Laboratory designers should consult the Radiation Safety Officer or licensee of the organization for information on protocols for handling radioisotopes which will influence selection of the hoods and accessories.

Perchloric acid hoods require interior lining and duct construction similar to that of radioisotope hoods. In addition perchloric acid hoods and ducts are installed with an internal sprinkler system that keeps explosive concentrations of perchloric acid from accumulating. Exhaust risers should be designed with as straight a vertical run as possible to the fan and stack. Perchloric acid hoods require regular flushing and maintenance. Safety officers should inform design engineers on these protocols, so the hoods will be specified and installed correctly.

California hoods are designed to be used with tall distillation apparatus or large

* *Code of Federal Regulations, Title 10,* Parts 1-50, Chapter 1, Nuclear Regulatory Commission.

Table 16.3 Typical Environmental Conditions Found in Specific Lab Types*

Lab type	HVAC Issues
General	Normal ventilation comfort standards and air cleanliness, some heat load from equipment; ASHRAE and ANSI have well-recognized ventilation standards for normal conditions
Biochemistry	High equipment heat load, filtration of supply and exhaust air for critical processes in biosafety cabinets, special hoods, and filtration for radioisotopes and perchloric acid
Physical chemistry and physics	High equipment heat load, very clean general supply air, local and spot exhaust sources, some of which may require filtration, critical sensitivity to air currents and vibration
Organic and inorganic	Large volumes of ventilation air, air currents
Theoretical	Normal ventilation comfort standards and air cleanliness, some heat load from computers and instruments

equipment that cannot fit within the interior chamber, or be accessed from the front opening of a standard fume hood. California hoods are sometimes called distillation hoods or incorrectly referred to as "walk-in" hoods. These hoods are not designed to be entered by researchers at any time. Simply, the interior chamber and sash opening(s), on one or more sides of the hood, are taller than in standard hoods. Work surfaces in California hoods may be as low as floor level. Work surface height in standard hoods is 37 in. above floor level.

In addition to standard and special fume hoods, local exhaust outlets may be required for special processes or to vent equipment. Many local exhaust devices are custom designed by health and safety professionals or experienced mechanical engineers. Please refer to Section III of this book for additional critical information on fume hoods and laboratory ventilation.

Table 16.3 gives a brief overview of ventilation issues that affect team laboratory design. Even where normal ventilation comfort standards prevail, laboratories should be negatively pressurized relative to corridors and nonlaboratory occupancies, and have careful design of air distribution near fume hoods. Biochemical labs have two ventilation requirements: cleanliness and containment, chemical and biological. To reduce contamination from sources outside a laboratory, researchers often ask that their labs have air pressurization positive to corridors and adjacent areas. This strategy may promote cleanliness in those labs, but it also creates unsafe, unhealthy conditions in public areas infused with potentially contaminated laboratory air. The solution is to provide a vestibule airlock or intermediate room that is negatively pressurized to both corridor and lab, and located between the clean lab and corridor. Leaky windows should be sealed to reduce infiltration of outside dirt. Laboratory occupants should wear clean lab coats and keep their outer garments and personal belongings in secured lockers

* Baum, DiBerardinis, American Chemical Society, 196th Annual Meeting, Committee on Chemical Safety Symposium, September, 1988.

outside the lab. Tacky mats placed at laboratory entries reduce particulate matter brought in on shoes. Laboratory protocols that promote personal cleanliness and reduce entry of outside sources of contaminants will improve performance of well-maintained lab ventilation systems.

Some physical chemistry and physics labs may have less need for containment than biochemical labs, but they require cleanliness and low vibration. Similar strategies for cleanliness may be used as for clean biochemical labs. For low vibration, low velocity distribution of supply air is effective. Air turbulence in ducts and from diffusers is a source of vibration, as is fluid turbulence in pipes to radiators, heating and cooling coils. All pipes and air handling units within or adjacent to vibration sensitive laboratories should be mounted with vibration isolation devices. Vibration sensitive areas should be located away from corridors, elevators, mechanical equipment rooms, or other areas with vibrating equipment nearby. The structure should be dense and rigid and isolated from the general structure if vibration is particularly critical.

Organic and inorganic team laboratories also benefit from low velocity air distribution, because of the large air volumes required. Equipment or processes that put out high heat loads should be removed from team laboratories and located in separately ventilated rooms. These rooms can be controlled better than large spaces for high ventilation rates and lower supply air temperature.

A caution on hazards in the laboratory would not be complete without understanding the impact of fire hazard on design of laboratories. Fire detection, alarm, and suppression systems in laboratories are mandatory in many jurisdictions.* Most health and safety professionals recommend that all new laboratory buildings, containing chemicals, should be sprinklered and that existing laboratory buildings which are not already sprinklered and which contain quantities of chemicals at or above the NFPA Class "C" category** should have automatic sprinkler systems installed during any major upgrade. Some mechanics and physics laboratories do not regularly use chemicals. They may contain other, less obvious fire hazards such as fine dusts or high voltage equipment that also justify installing sprinklers. Sprinklering and other appropriate fire suppression and detection systems do more that any other technology to reduce loss of life and real property. If water from sprinklers may cause damage to expensive equipment, chemical suppressant media may be an appropriate method. Smoke and fire detection devices that reduce incidence of false alarms in laboratories are highly recommended. One type is a detector that senses obscuration of the air by smoke. It gives very early warning more accurately than ion type detectors. In some applications, detectors that measure rate-of-rise of ambient temperatures may be preferred. There are systems which can detect over-heating at specific pieces of equipment or processes. Safety professionals can assist in system selection and specification.

* NFPA 13, p. 1 - 106.

** NFPA 45, Appendix B and NFPA 704, p. 7.

Local zoning and building codes as well as the National Fire Protection Association (NFPA) set detailed standards for fire resistive construction of laboratories and on fire separation of labs with other building uses. *NFPA 101, Life Safety Code,* deals with general building egress and construction issues. *NFPA 45, Laboratories Using Chemicals,* is specific to laboratory functions. By *NFPA 45* definition, a team lab including support rooms, may be regarded as a single laboratory unit. *NFPA 45* classifies the hazard of the laboratory unit, regulates the allowable quantities of flammable and combustible chemicals within a laboratory unit, regulates maximum allowable lab area according to construction type, and requires fire separation of the laboratory unit from other occupied areas. In *NFPA 45*, "Table 1.3", quantities of each class of flammable and combustible materials contained in the laboratory unit determine the hazard classification of the unit to be A, high hazard, B, intermediate hazard, or C, low hazard. Flash point and boiling point determine the class of flammable and combustible materials Class I, IC, IB, IA, II, III, IIIA, IIIB, and IV.[12] Quantities of flammable and combustible materials allowed in a laboratory unit are regulated not only by overall quantities but also by gallons per unit area, by storage container and method, and by level of sprinklering. Fire Hazard Class A allows laboratory units up to 5000 ft^2 in unsprinklered Construction Type I (noncombustible) and II (limited combustible) buildings. This limit doubles to 10,000 ft^2 if the laboratory building is fully sprinklered. For Fire Hazard Classes B and C, area and construction type requirements are less stringent. As shown in *NFPA 45*, "Table 1.4", required fire separations of laboratory units from other laboratory units and nonlaboratory occupancies vary according to hazard and level of fire protection. Another reference is NFPA 45, Table B.4, "Fire Resistance Requirements for Type I through Type IV Construction".

Safety professionals should advise the owner and designers under which Fire Hazard Classification the laboratories will operate. Laboratory designers should insist on written confirmation of this information early in the design process. However, more often than not, designers do not ask for or are not given this directive. During design if building occupants are not known and the owner is uninformed or noncommital, the Fire Hazard Classification of the laboratory unit is left for the design team and the construction budget to determine. After occupancy, the Fire Marshal or the building insurer, with the cooperation of safety professionals, regulate the quantities of chemicals in the lab to comply with the Fire Hazard Classification that was actually constructed. This default position may not meet the needs of some organizations. (Class A Fire Hazard classification requires very special design considerations to protect life and property. Laboratories, constructed by general standards, do not meet the strict fire safety requirements of Class A.) Planning in advance for a reasonable density of chemical supplies for team laboratory units is a preferred strategy. Size and distribution of chemical storerooms strongly affects amounts of chemicals stockpiled in laboratories.* Underwriter's Laboratory and Factory Mutual labeled storage cabi-

* Young, *Improving Safety in the Chemical Laboratory*, p. 205 - 218.

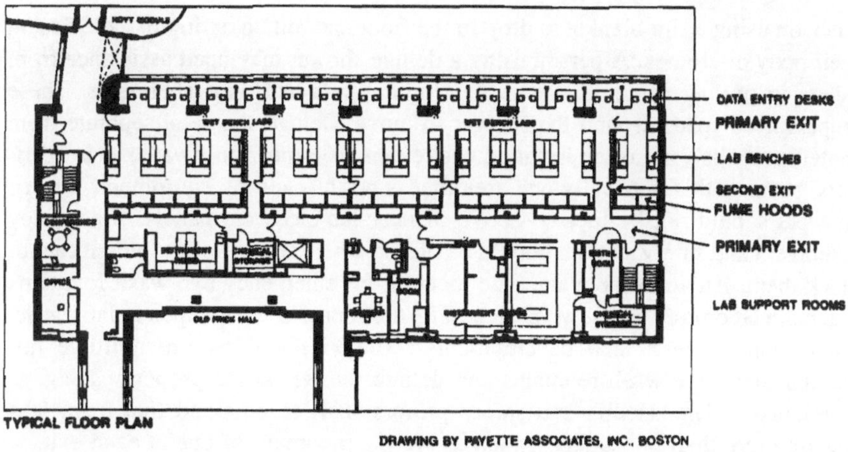

Figure 16.2. Example of hazard zoning in Frick Chemistry Laboratory, Princeton University.

nets for flammable liquids are suitable for storage of limited quantities of these chemicals within team laboratories.* However, equipment, materials, or processes that pose higher than normal hazards should not be located within a team laboratory. They should be put in separate rooms with appropriate safety considerations for fire safety, alarms, local exhaust, ventilation, containment, emergency shutoffs, and other safety equipment.

Emergency equipment such as portable fire extinguishers, fire blanket, emergency eye wash and safety deluge showers , and spill kits should be provided in each lab. The number and location of each will depend on the size of the laboratory and the number of people. Safety engineers recommend that there be no more than a 20 ft, a 5 to 6 sec walk, to an eye wash fountain and maximum of 50 ft to a deluge shower.** In evaluating a lab layout, actually check the time it will take to reach emergency equipment. The distances may be within the prescribed limits, but if the location is not clear or easy to get to, utility of emergency equipment is compromised. Certain labs may need respiratory assist equipment also.

A laboratory "safety station" can be designed to hold and conveniently organize most emergency and safety equipment. A safety station should include safety glasses, latex and other protective gloves, portable fire extinguisher, fire blanket, and tack board for safety information and notices. It may also include an emergency eye wash fountain and deluge shower and respiratory assist equipment. In new or renovated labs the safety station should be in a consistent location from lab to lab, such as near exit doorways. A safety station can be developed at an unobstructed wall area, or in a tall laboratory cupboard modified to serve the purpose, that faces a major laboratory aisle. The aisle provides open floor area for

* NFPA 30, p. 24 - 31.

** Weaver, Britt, "Criteria for Effective Eyewashes and Safety Showers", p. 38 - 54.

a person using a fire blanket to drop to the floor and roll, to extinguish the fire on their body or clothes. A person using a deluge shower may need assistance from others in pulling off outer layers of clothing splashed with chemicals. These helpers need space around the shower to move. Deluge showers, separate from "safety stations", are often installed above exit aisles near doorways, or in corridors, because this is usually one area that is not blocked by equipment. Deluge showers should not be located above or near lab electrical panels. Emergency equipment and safety stations should be placed where scientific equipment cannot block them. An acceptable alternate location for emergency eye wash fountains is at main laboratory sinks, which generally are central and kept open to lab traffic. Large team research and development laboratories will require multiple fire extinguishers, eye wash fountains and deluge showers at the proper spacing, as mentioned before. Health and safety professionals recommend that team labs contain more than one safety station as well, a minimum of one at each exit.

Apply Principles of Hazard Zoning

The first principle of hazard zoning is to locate the highest hazards away from the primary egress of the laboratory. The second principle is to locate the activities of lowest hazard along or near the primary egress. Between the highest to the lowest hazard zones, activities of moderate hazard to lesser hazard will be located. Figure 16.2 illustrates a good example of this zoning in organic chemistry laboratories designed by Payette Associates, Boston and constructed at the Frick Chemistry Building, Princeton University.

Please observe that the primary egress is between data entry desks and the dry ends of benches. Chemical fume hoods and secondary egress aisle are on the opposite side of the lab module. Lab sinks, wet benches, distillation racks, and reagent shelves are between. Each standard module is designed in this manner. On the other side of the structure and across the fire rated egress pathway are special laboratories and support rooms. These include chemical storage rooms, distillation labs, cold rooms, and special equipment labs for glove boxes and lasers. This zone contains activities, apparatus, or materials that pose higher hazards or ventilation conditions that cannot be safely contained in the open team laboratory module. One end of each laboratory floor contains a conference room or lounge, private and group offices for laboratory personnel. These are safe, appropriate spaces for quiet study and conversation, where attention is not focused on the equipment or chemical reactions. Lab personnel can eat and drink there with lower risk to their health than in the lab. Many possible variations in layout can achieve good hazard zoning. It costs no more to organize the laboratory based on these principles. Locating chemical fume hoods next to the exit doorways in labs is one common violation of hazard zoning principles. This condition not only disrupts proper air flows into the fume hoods, it poses unnecessary risks to persons leaving laboratories under normal but especially emergency situations.

Plan for Future Expansion to Avoid Overcrowding

Laboratory buildings have a structural grid of columns or bearing walls supporting floors or a roof above. Ideally, module dimensions coincide with the grid of the structure. Module dimensions are usually determined very early in the design process, early enough to determine a compatible structural grid. (Renovation of existing structures may require adjustments to module dimensions because of inconvenient or unsafe column locations.) The standard acceptable range of a single module width, centerline to centerline, is 10 ft 6 in. to 12 ft. Double modules are 21 to 24 ft. A module grid of only 10 ft is not recommended because this width does not allow a minimum clear aisle width of 5 ft under certain conditions. When both sides of a single module are lined with large pieces of equipment, such as a fume hood, centrifuges, or cold boxes, which are 3 ft in depth or require clearance in the rear for air circulation, the aisle may diminish to under 4 ft. Module widths over 12 ft are less efficient and encourage researchers to fill the wide aisle with equipment, gas cylinders, and tables, which clutter and reduce the effective aisle width below a safe minimum.

The length of a module is more variable, within limits, and less critical to safety. Building codes may define the maximum distance, such as 50 ft, from the furthest point of any room to the centerline of the nearest egress doorway. The maximum length of a laboratory may be based on that code requirement. However, in current laboratory practice, module lengths vary from 20 to 35 ft. Single module areas, within dimensional ranges given here, range between 210 to 420 ft^2. According to laboratory type, function, and desired density, a single module may accomodate 1 or 2 persons. Double modules may accomodate 2 to 4 persons. Team laboratories, as noted in the opening paragraph, may extend many modules and accommodate proportionately more researchers. Figure 16.3 illustrates module dimensions, area, and proportions.

In order to estimate an acceptable population density in a laboratory the following chart can be used as a guide. The following chart (Table 16.4) reflects area utilization in several typical laboratory types. These areas are only intended as a preliminary and general guide, not a standard. Personnel occupying this area are assumed to be trained in laboratory safety and emergency protocols. These figures do not apply to standards for teaching laboratory use. The actual area requirements of a laboratory will be based on the specific quantity and toxicity of chemicals and materials used, processes contained therein, and the size and type of equipment needed. The areas also do not necessarily imply a change in module area from one laboratory type to another. It indicates the research population that can be expected to safely occupy a specified area. In deciding on the size and dimensions of a module, it is prudent to select the more conservative area guide for the wide range of activities projected for the facility. For instance, if the laboratory types for a new facility include organic chemistry along with biochemical and general or analytic chemistry, the area per researcher for biochemical labs and fume hood per researcher guides should be selected for the lab module design.

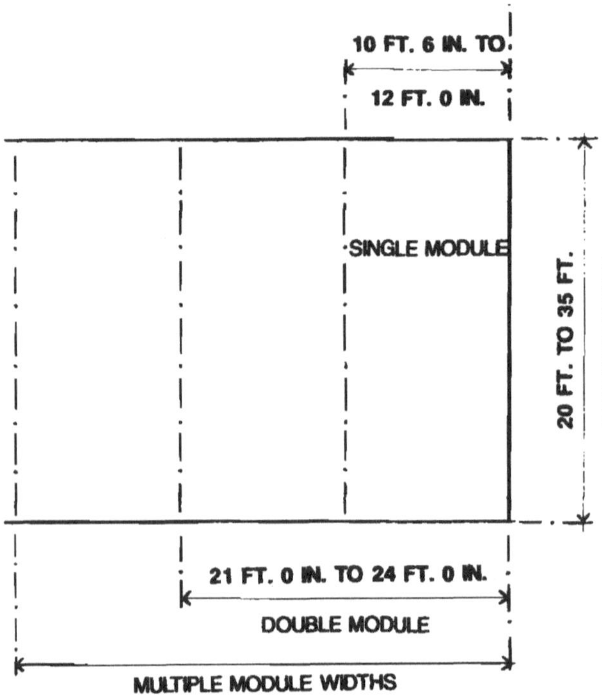

Figure 16.3. Module diagram.

4. CONCLUSION

Team laboratories present several serious design problems:

They are more open.
There are more people, movement, and activities.
Accidents are more difficult to control and contain; more people are exposed.
They are more difficult to ventilate and require larger volumes of air.
It is more difficult to control air distribution and to maintain heating and cooling loads.

Despite these problems team laboratories can be designed to provide stimulating and safe environments for scientific research and development work. Over time a team lab can be much more flexible than a series of individual's laboratories. Territory and ownership issues diminish; team labs promote sharing and exchange of ideas and equipment. This chapter has addressed some of the issues of design and construction of team laboratories and the participation of health and

Table 16.4 Typical Laboratory Area Utilization Sample

Lab type	L.F. fume hood/res	S.F. lab bench/res	S.F. equip area/res	S.F. office use/res	Total area/ researcher (S.F.)
General	3	145	50	35	260
Biochemistry	3–6	150	120	35	335
Physical	3	50	140	65	285
Organic	6–12	160	60	45	315
Inorganic	5–6	145	50	35	275
Theoretical	0–3	100	25	100	225

Note: RES. is researcher, L.F. denotes length in linear feet, S.F. is net floor area in square feet

safety professionals in the process. Safety depends not only on these, but management's policies and individuals' decisions that promote a safe environment. Safe laboratory design sets the stage for this commitment.

References

1. ANSI Z53.1-1979, "Safety Color Code for Marking Physical Hazards", American National Standards Institute, Inc., New York, 1979.
2. ANSI/EIA-359-A-1984, " EIA Standard Colors for Color Identification and Coding", Electronic Industries Association, Washington D.C., 1985.
3. *Code of Federal Regulations, Title 10: Energy*, Parts 1-50, Chapter 1, Nuclear Regulatory Commission, Office of the Federal Register, January, 1989.
4. "Criteria for Effective Eyewashes and Safety Showers", A. Weaver and K. Britt, *Professional Safety*, v. 22, 1977.
5. *Guidelines for Laboratory Design: Health and Safety Considerations,* DiBerardinis, Baum, First, Gatwood, Groden, Seth, John Wiley & Sons, Inc., New York, 1987.
6. *Improving Safety in the Chemical Laboratory*, Jay Young, Ed., John Wiley & Sons, Inc., New York, 1987.
7. "Laboratory Fume Hood Specifications and Performance Testing Requirements", Environmental Medical Service, Massachusetts Institute of Technology, 1979.
8. *Method of Testing Laboratory Fume Hoods*, ANSI/ASHRAE Standard 110-1985, American Society of Heating, Refrigeration, and Air Conditioning Engineers, Atlanta, 1985.
9. *NFPA 10*, "Portable Fire Extinguishers".
10. *NFPA 13*, "Sprinkler Systems".
11. *NFPA 30*, "Flammable and Combustible Liquid Code".

12. *NFPA 45*, "Fire Protection for Laboratories Using Chemicals".
13. *NFPA 101*, "Life Safety Code".
14. National Fire Protection Association, Quincy, Massachusetts, 1985.

CHAPTER 17

Safe Handling of Hazardous Chemicals in a Chemical Containment Laboratory—Design Requirements of a Chemical Containment Laboratory

Richard A. Smith, Lawrence H. Keith, Richard L. Trammell, Douglas B. Walters, and Andrew T. Prokopetz

1. INTRODUCTION

Research which involves handling large numbers of varied hazardous materials bears with it the potential for personnel exposure and environmental contamination. The ability to safely and efficiently handle milligram to kilogram quantities of such materials has been designed and built into two Chemical Containment Laboratories (CCLs). Major functions of the CCLs include support of a chemical repository for the National Toxicology Program (NTP), dioxin syntheses, separations and purifications, environmental and industrial hygiene standards preparation, dosing and mixing of animal feeds, and protective clothing.

2. CHEMICAL CONTAINMENT LABORATORIES

The basic guilding philosophies for the design and operation of a CCL have been stated by Harless as:[1]

- Protect the personnel

- Protect the environment
- Protect the chemicals
- Protect the facility

Combining these philosophies with the realities of construction and maintenance costs, productivity, and worker morale will result in a versatile facility which ensures personnel and environmental protection in a cost-effective manner.

The basic characteristics that should be incorporated into the design of any modern CCL are listed below.

- Engineering controls and work practices should serve as the primary protection of both personnel and the environment.
- Chemical protective clothing is relied upon as a secondary, redundant line of defense.
- The facility should always operate at various negative atmospheric pressures rather than ambient pressure and, of course, never at positive pressures.
- Handling and storage of chemicals must be conducted in well-defined and controlled areas of the building.
- Access to the CCL must be limited.
- All effluents and wastes must be controlled and monitored.
- All ventilation systems should be redundant and be provided with emergency back-up systems if possible.
- Emergency electricity to power the ventilation system, lights, and refrigeration units should be provided.
- Personal injury (manually activated), fire, and mechanical failure alarms must operate throughout the facility.
- Utilities such as nitrogen and argon gases, vacuum, and liquid nitrogen should be built into the facility if possible.

3. SITE LOCATIONS

Site selections for CCLs are critical and are affected by accessibility, security of non-CCL personnel, building and zoning codes, and construction constraints. One of the facilities was attached to an existing building by placing storage and accessways between the two structures. This gave the advantage of proximity with the protection of separateness — a nearly ideal situation. However, the other CCL was constructed as part of a larger laboratory facility. Although not as ideal as a separate facility, this situation is likely to be the more common of the two.

4. WORKSPACE DESIGN

Radian's Austin, Texas facility is an example of the integration of the buiding

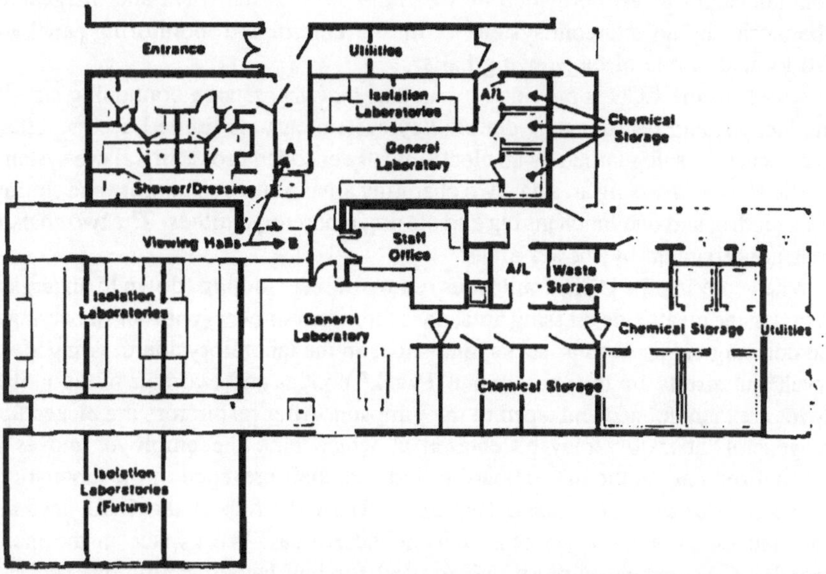

Figure 17.1. Radian chemical containment laboratory (CCL) floorplan.

philosophies for design of a modern CCL. It occupies a separate 5000 ft^2 building one-and-a-half stories high; the main floor of the building houses the showers, dressing rooms, an airlock, and laboratory facilities, and the upper half-floor contains most of the utilities and mechanical equipment serving the laboratory. Interior walls are gypsum wallboard covered with three coats of epoxy polymer-based paint. Floors are covered with vinyl sheet flooring, and seams are sealed by thermal welding. The flooring is coved 6 in. onto wall surfaces in the laboratory portions of the building as recommended by Harless et al.[2] The entire laboratory is operated at pressures negative to the atmosphere and there are pressure differentials from room to room, as discussed later.

The operational elements incorporated into this facility are

- External access
- Shower/dressing
- General laboratories
- Isolation laboratories
- Chemical storage
- Waste storage
- Office

All laboratory workspaces can be viewed from one of two viewing halls (A or B, as identified in Figure 17.1). This important safety feature allows personnel external to the facility to have visual contact in the event of emergency situations.

Communication is also provided between the viewing hallways and the general laboratories by an intercom system. A master control and monitoring panel are also located in one of the viewing halls.

Access to the CCL is provided by an *external access* area containing sign-in log, lockers, and restrooms. A double-key system ensures limited access. After-hours access is safeguarded by an electronically encoded proximity alarm system.

The *shower/dressing* area has two changing areas: one for changing and storing street clothes and one for changing and storing laboratory clothes. The two change areas are separated by shower areas.

When entering the CCL, employees remove street clothing (down to undergarments) in the clean-side dressing area, pass through a shower (without showering), and don surgical scrubsuits, socks, and shoes in the laboratoryside dressing area. Tyvek® jumpsuits are then put on and Tyvek® booties and two pair of dissimilar gloves are then donned and taped to the jumpsuit. After respirators are picked up, the general laboratory may be entered at which time the employee moves a personalized peg on the in-out board to indicate their presence in the laboratory.

The *general laboratory* areas (Figure 17.2) are the hub of the CCL since all other laboratory areas are accessed from these areas. Workspace in the main laboratories is general purpose and is used for low hazard operations such as instrumental analysis, and ampule sealing and packaging. Much of the safety equipment, including emergency showers, eyewashes, first aid kits, and fire extinguishers, is kept in this area.

The *isolation laboratories* are equipped to support specific chemical handling activities. Each contains a fume hood and is isolated from the general laboratories by glass walls and sliding glass doors. All operations involving hazardous materials (other than storage) are performed in the isolation laboratories. Additionally, open vessels of hazardous materials are allowed only within confines of the fume hoods. Respiratory protection is required in the isolation laboratories at all times.

Chemical storage areas have a high potential for contamination and, thus, are isolated from all areas and vented mechanically to the outside. Ten walk-in chemical storage areas are housed in the CCL, including four at ambient temperature, three at 5°C, and three at –20°C. All chemical storage areas have explosion-proof fixtures. A subfloor vault is provided for certain extremely hazardous materials and an alarmed controlled substance safe is also utilized.

The CCL *waste storage* area is used to accumulate, segregate, package, and temporarily store laboratory wastes prior to their disposal.

The *office* area provides workspace for data collection and reduction, reference materials, and communication. It also provides a space where laboratory personnel can relax without having to undergo the procedures required to exit the CCL. A computer terminal which is linked to the hazardous chemical materials database is housed in this area.

Two secondary *external access* points are provided to allow movement of chemicals, supplies, and equipment into the CCL. Both are connected via air-locks to avoid possible environmental contamination.

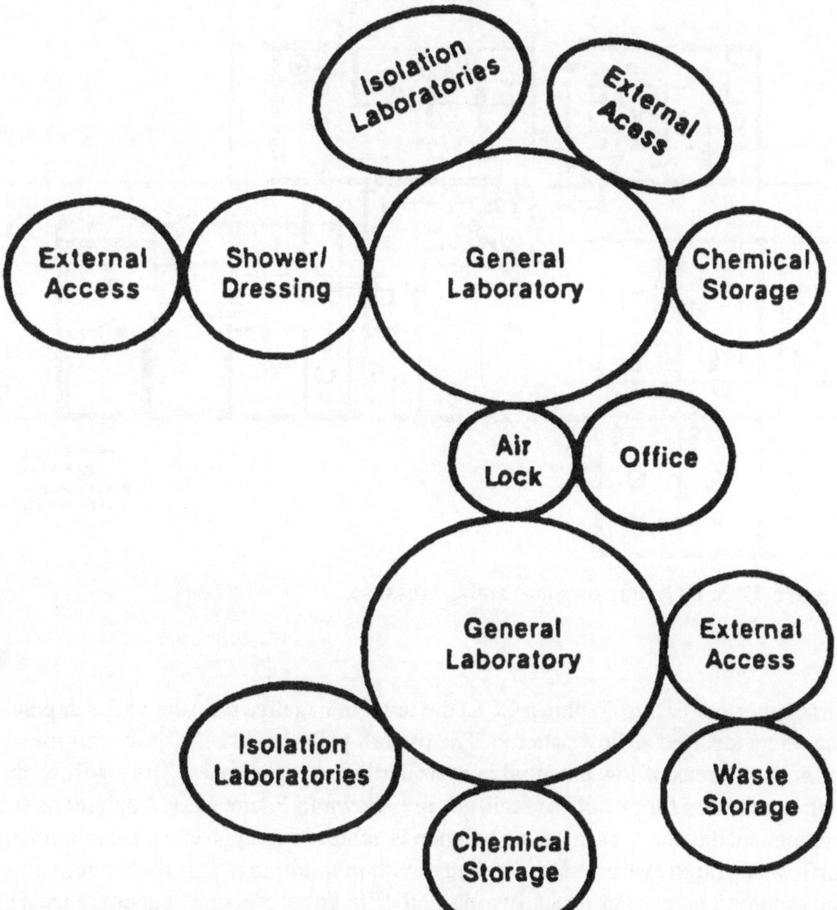

Figure 17.2. Workspace elements.

5. MECHANICAL/UTILITIES DESIGN

The four major categories of mechanical elements in a modern CCL consist of:

- Atmosphere control and monitoring
- Effluent control
- Utilities
- Critical systems control and monitoring[1]

Atmosphere control within a CCL is provided by operating at a negative

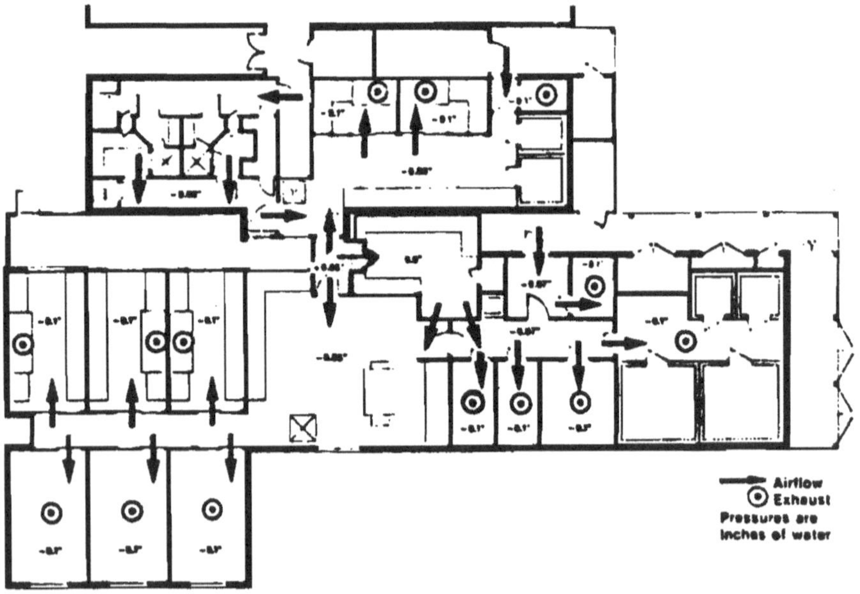

Figure 17.3. CCL airflows and static pressure.

atmospheric pressure. Within a CCL, the level of negative pressure varies depending on the desired airflow patterns. The overall plan should be to ensure air movement from areas of low potential contamination to higher ones. The air-flow and static pressures for one of the facilities are shown in Figure 17.3. Attainment and maintenance of such a critical air balance is achieved through exact calculation of airflows through every vent and exhaust with monitoring to ensure they remain at the specified levels. Permanently mounted differential pressure gauges are used to monitor pressures between adjacent areas. Duct flowrates are routinely measured to ensure proper operation.

Effluent control is a critical aspect of the operation of any CCL due to the type and amounts of exhausted air, wastewater, and laboratory wastes. The exhausted air and wastewater must be managed using engineering controls while laboratory wastes exit the facility packaged for disposal.

All exhausted air passes through a filtration system. The rough particulate prefilter traps most particulates, and is followed by an absolute filter. Next, an activated carbon filter absorbs most organic vapors and a terminal high efficiency particulate absolute (HEPA) filter collects most aerosols and fine particulates. Pressure drop across *each* filter bank is monitored with differential pressure gauges mounted on a control panel. These gauges are used as a measurement of filter life. When the pressure drop across a filter is approximately three times that when new, it is replaced. Filters are changed on a routine basis using a "bag-out" process in which contaminated filters are withdrawn into heavy plastic bags for disposal.

An auxiliary exhaust fan backs up the main exhaust fan in the event of mechanical failure. This is an important aspect of the backup philosophy used with critical units.

Wastewater treatment systems are based on an accepted design as described by Nony et al., and are composed of a collection tank, pump, particulate filters, and chemical adsorbents.[3] Wastewater is collected in a holding tank, then pumped through successively finer particulate filters (40, 25, and 5 μm) then through adsorbents (charcoal and XAD-2), and then finally into a sanitary sewer. Samples are periodically analyzed to ensure adequate treatment.

Utilities serving most CCLs should include: hot and cold water, deionized water, compressed air, natural gas, vacuum, argon, nitrogen, breathing air, and electricity. All pipes and conduits carrying these services in the subject facilities were constructed in the overhead chase space whenever possible, and their entrance into the laboratory was made through the ceiling or walls as required. Whenever an inner wall was traversed, the area around the incoming pipe or conduit was completely sealed, first with polyurethane foam and then with a silicone polymer sealant. This is done to assure maximum isolation of the laboratory atmosphere and so that the negative pressure regulation can be precisely maintained.

Critical systems control should be a major factor in the design and operation of a CCL since a maximum level of protection is required at all times. This requires that the facility operate at design specifications 24-h a day, every day; that operational parameters are not significantly affected by mechanical failures; and that the physical structure is preserved. The following critical systems are key to this goal:

- Electrical power
- Air exhaust
- Fire control
- Effluent treatment
- Breathing air
- Chemical storage refrigeration

Each critical system must be backed up with a redundant system whose startup is triggered by a failure. In addition to starting the back-up system, the failure monitoring system should activate audible and visible alarms. Systems control, monitoring, and alarm functions should be centralized in a master control panel if possible.

6. STRUCTURAL DESIGN

Generally, structural design is within the domain of the architect; however, due to the nature of a CCL, there are three areas of special concern:

- Floorplan design
- Utilities access
- Equipment placement and loadings

Each of these areas entail aspects which, if not properly addressed, may result in clean area contamination, non-CCL employee exposure, environmental contamination, and/or surface contamination buildup over a period of time.

Floorplan design is critical in ensuring proper access, traffic flow, and utilization. Because of the high construction costs of any CCL, facilities should be designed with expansion in mind. The following elements incorporated into the basic structure will greatly facilitate future expansion with minimal cost at the time of the initial construction:

- Utilities may be sized and located for ready connection and support of additions
- The CCL shape should allow for expansion
- Means of access to additions may be built-in
- The foundation and other structural elements should be compatible with future expansion areas

Utilities access in a CCL require significant planning because all laboratory surfaces must be sealed (no false ceilings) and utilities should be placed inside permanent ceiling or wall spaces.

Our approach was to place the utilities in an 8-ft high interstitial space above the CCL. Although expensive, this approach provides much better access than using utility alleys on the same level as the CCL.

Equipment placement and loadings, particularly air handling, is a significant planning concern because of the sheer size and mass of the equipment involved. The roofs may be virtually covered with air-handling equipment.

7. EQUIPMENT DESIGN

The major types of equipment in a CCL include laboratory furniture, fume and canopy hoods, safety equipment, and special operations equipment, such as chemical storage units. All equipment surfaces must be of material which can be readily decontaminated and, thus, solvent resistant. Fume hoods and cabinetry should be stainless steel, and epoxy resin is recommended for counter-top surfaces. CCL fume hoods should incorporate a single-unit stainless steel interior with welded seams, coved corners, and a lip on the front to prevent runout of spilled liquids. Recommended mean face velocity of each fume hood is 100 ± 20 ft/min. Monthly audits insure the optimum velocity.

Chemical storage units may include solvent cabinets, ambient and subambient temperature storage areas, and vaults. The size of each area will be dictated by the scope of operations. Location of each unit should be determined by considering personnel and traffic flow.

Major safety equipment planning is critical from both the selection and placement aspects. Traffic flow and proximity to hazardous areas must be carefully considered in placement of all safety equipment.

8. SAFETY DESIGN

Safety design has been discussed throughout this article; indeed, safety considerations must be the primary driving forces behind the design and construction of any CCL. There are some specific areas which are considered as purely safety matters: visibility, access control, fire protection, communication, alarm systems, and information access.

Visibility is critical in providing a safe work environment in any CCL. Internal and external visibility are ensured by utilizing glass for a large number of the walls. Internal visibility allows early recognition of problem situations by coworkers who must monitor each other and respond quickly in case of an emergency condition. Good external visibility is essential in emergencies to allow assessment and direction by persons from a safe vantage point.

CCL *access* is limited to those who are medically monitored and qualified to work therein. Entrance should be gained via only one locked door with keys distributed only to those personnel who are qualified to enter. Other doors (for example, fire exits and loading dock doors) should be locked from inside the CCL. Intrusion alarms will ensure that access in off-hours is by qualified personnel only.

Fire protection can be maintained by a network of ionization and thermal detectors which activate extensive sprinkler systems or Halon™ within a CCL.

Communication should be provided by two-way intercoms which allow verbal contact with people outside the CCL. The intercom points should be located such that visual contact can also be maintained. All transmissions should be heard at all stations in case of requests for assistance.

A number of *alarms* must be used in any CCL to ensure that the CCL staff and outside personnel are apprised of emergency conditions. Mechanical failure alarms may be linked with all major mechanical devices to monitor their status continuously. Personal injury alarm activators should be placed throughout the facility. These alarms should be easily distinguishable from one another and sound in the CCL, other laboratories, other offices, and selected other buildings. We recommend that the alarm system be monitored continuously by an outside contractor with notification and written action plans in place.

Information access is a key means of ensuring worker protection by familiarization with the properties, hazards, and emergency responses associated with the materials handled. Radian maintains a computer terminal within the laboratory itself which has immediate emergency access to the NTP database which houses this information. An emergency access selection can display first aid, fire fighting, and toxicological, chemical, and physical properties of any of the 2000 chemicals in the NTP chemical repository within seconds. This option can be implemented by using one of the several computerized MSDS software packages available or, in the case where few chemicals are used, with printed copies of MSDSs.

9. CONCLUSION

Designing a CCL is a dynamic process because of the growing body of knowledge relating not only to the hazardous nature of materials, but also to personnel protection, effluent control, and materials handling. Radian Corporation has designed, built, and maintained two CCLs to allow for safe and efficient handling of a wide variety of hazardous and potentially hazardous substances. Each aspect of the facilities was designed to afford the utmost in personnel and environment protection.

While not necessarily meeting all the needs of future CCLs that will be designed or built, the design principles discussed here are basic to almost any CCL. They have been validated with 8 years of intensive operation and, thus, are "tried and true" having withstood the test of time.

While seemingly stringent in many aspects, most of the design criteria recommended are necessary if one is to provide maximum safety for employees who have to work with hazardous chemicals. Needless to say, the design and construction of CCLs is expensive. But short cuts are unacceptable to the worker, the public, and to professionals in chemical health and safety. Besides, liability considerations quickly make short cuts cost prohibitive. Therefore, a carefully thought out design which incorporates all the workplace needs while maintaining all of the safety and engineering controls is the first step in planning a CCL either as a modified laboratory within an existing structure or as a new stand alone building.

References

1. Harless, J. M., "Components in the Design of a Hazardous Chemicals Handling Facility," in *Health and Safety for Toxicity Testing*, (Butterworth Publishers, Stoneham, MA; pp. 45-72, 1984). Walters, D. B. and Jameson, C. W., Eds.
2. Harless, J. M., K. E. Baxter, L. H. Keith, and D. B. Walters, "Design and Operation of a Hazardous Materials Laboratory", in D. B. Walters, Ed.: Safe Handling of Chemical Carcinogens, Mutagens, Teratogens, and High Toxic Substances, (Ann Arbor Science, Ann Arbor, MI 1980).
3. Nony, C. R., E. J. Treglown, and M. C. Bowman, "Removal of Trace Levels of 2-Acetylaminofluorene (2-AAF) from Wastewater", *Sci. Total Environ.*, 4, 155-163, (1975).

Section V: Working Together to Design Safe Laboratories

CHAPTER 18

Where Laboratory Design Projects Go Right or Wrong

Norman V. Steere

Laboratory design projects are more likely to go right if the owner recognizes some basic problems and makes some important decisions at an early stage, *before* a budget is established or a design firm is selected. Other problems can be prevented during the design process.

One serious problem which can cause a project to "go wrong" is to have a design and construction budget which is too low to cover the costs of the laboratory which the owner needs and expects to build. Many preliminary budgets are set unrealistically low because someone does not recognize that technical buildings are much more expensive per square foot of usable space than most other kinds of buildings. For example, the costs of mechanical equipment for efficient laboratory ventilation are far higher than those for ventilation of warehouses or offices since greater treatment must be given to the air supplied to laboratories and the air removed cannot be recirculated. We have seen projects in which important space had to be eliminated late in the design process because the initial budget did not reflect actual construction costs.

Another problem that can cause a project to go wrong is a lack of information for the design team on the functions that the laboratory must accomplish. New or remodeled laboratories are more likely to meet the needs of laboratory users if they are given an opportunity and the time for participation in the planning and design process. Initially, all laboratory users and occupants should participate in the definition of the functions that the building is intended to accomplish. After the design firm has been selected, users' representatives should be given time and responsibility for continuing participation throughout the design process.

There can be problems in designing suitable facilities if the owner uses terms not familiar to the architect or uses them with different connotations. For example, some of the terms which can be unclear are "chemical drum storage", "solvent storage", and "hazardous chemical storage". These terms do not include design-necessary definitions of sizes of containers, acceptable storage configurations, type of solvents which may be stored, whether storage areas will also be used for dispensing, whether dispensing will be of limited or large quantities, and what hazards will require special design such as ventilation.

To facilitate design, construction, and adaptability of the laboratory for future uses, the users should recognize and agree before the project begins that the design of the structure, utilities, and most spaces should be modular and that idiosyncratic requirements should be kept to an absolute minimum.

An important requirement that should be recognized early in the design process is that the design of the structure must provide adequate vertical service chases and space between floors so that mechanical systems such as ventilation can be installed, maintained, and operated efficiently. We have seen buildings in which the space in utility areas was so limited that ventilation systems could not perform efficiently and equipment could not be reached for service.

Many projects fail to be as good as they could be as a result of inadequate space being provided for movement and for storage. Corridors that are just wide enough for two people to pass each other may become constricted if there is no allowance for temporary "parking" of carts used for deliveries or pickups, or if laboratory occupants are allowed to use the corridors for storage. If corridors have drinking fountains, they should not obstruct movement of pedestrians or people in wheelchairs. If emergency equipment such as safety showers, spill kits, and breathing masks are located in corridors, adequate space should be provided for access and maintenance.

Some potential problems in corridors and work areas can be anticipated by looking at present areas that are crowded or obstructed. See what storage or working space should be provided for delivered supplies, glassware set out for cleaning, copy machines, break areas, CRT (terminals), printers, and waste stored for pickup, analysis, or treatment.

(As part of the "Owner's Manual" recommended for any new laboratory, it should be clear that corridors are the equivalent of the "public right-of-way" and laboratory occupants are not permitted to use corridors for any type of storage.)

If long or large equipment will have to be brought into the laboratory, the corridor should be wide enough to allow for movement and turning of such equipment. The width of doors, or doors plus inactive leafs, should be adequate to allow movement of the largest equipment which is likely to be installed.

Corridors may also become a problem if rough floor surfaces impede safe movement of carts or personnel. Use of decorative floor surfaces such as brick and architectural tile should be limited to areas where it will be safe.

Separate service corridors, built as an effective means of distributing utilities and providing adaptability for changes, can also be used for movement of materials which may otherwise conflict with pedestrian traffic. To prevent misuse or crowding of service corridors, a decision is needed on what equipment and storage will be located in the corridors as part of the design. Since service corridors may serve as alternate exit routes from laboratories, storage of hazardous equipment and materials should not be located in such corridors.

A laboratory project will work better if the users and designers think of maintenance and service needs and plan accordingly. In some cases future problems can be minimized by providing adequate access to equipment and an adequate number of shut-off valves. For example, a water leak in one large building required shutting off the water to ten floors of the building to allow repair of the leak.

Laboratory design projects often go wrong because of deficiencies in the laboratory ventilation system. Some common examples are exhaust outlets which do not direct the exhaust up and away from windows and fresh air intakes, high-velocity air supply devices which interfere with hood performance, and failure to provide exhaust ventilation for all sources of irritating, corrosive, and toxic vapors and aerosols.

To prevent cluttering of laboratory work areas, adequate space is needed for storage of chemicals, glassware, supplies, wastes, and infrequently used equipment. Adequate space is also needed to allow laboratory equipment to be serviced and repaired.

Following past design practices often leads to wasteful planning and safety problems. For example, the number of service outlets and cup sinks provided in new laboratories should be based on an assessment of needs, and not merely be a replication of a previous design. Outlets and cup sinks that are not likely to be used are a waste of money and space. Sink and floor drain traps that are not used will dry out and allow odors and gases from drain lines to escape into work areas.

Although two-dimensional drawings in a reduced scale will be used to build or remodel a laboratory, such drawings may not let laboratory people know what to expect. Many people find it difficult or impossible to visualize a new work area from design drawings, and it may be very helpful to lay out important parts of the laboratory at full scale in an available large area such as an auditorium or warehouse; furniture can be simulated with tables or boxes. In some cases, the expense of a full-scale mockup may be justified to communicate with laboratory users and with contractors who will bid on construction. Mockups are particularly important where utility services must be coordinated to avoid conflict that may result in more expensive construction or less efficient operation.

Serious problems in construction or in maintenance can result if the engineering designs are not properly coordinated by the architect because of lack of time or money. In many cases it will be very desirable to use a computer aided design (CAD) system to superimpose drawings of the structure and all utilities and ventilation ductwork. As one example, if the drawings do not coordinate the location of electrical panels and major ductwork and the electrical panel are installed first, changes in major ductwork may *greatly* decrease the performance of the ventilation system. Although there may be some initial added cost for coordinating the drawings, the costs of not coordinating them may be much greater.

Another construction problem may arise from specifications for custom items that allow bids on the item or its "equal". This creates confusion among bidders and difficulty in determining what is "equal". When three bids are required by law, as is usually the case for construction of public buildings, it may be difficult to compare three so-called equivalents to the same item and make the best choice. It will generally be better to specify standard items so that all can bid on the same item. Two other possible options are to detail the custom item in exquisite detail, or to purchase the item separately later.

The best ways to get laboratory design projects to go right are to start with a realistic budget for adequate space, develop a thorough description of the functions the new laboratory must perform, and maintain good communication between laboratory users and the design team.

Index

Index